From Bessel to Multi-Index Mittag–Leffler Functions

Enumerable Families, Series in them and Convergence

From Bessel to Multi-Index Mittag–Leffler Functions

Enumerable Families, Series in them and Convergence

Jordanka Paneva-Konovska

Technical University of Sofia, Bulgaria
&
IMI - Bulgarian Academy of Sciences, Bulgaria

 **World Scientific**

EW JERSEY · LONDON · SINGAPORE · BEIJING · SHANGHAI · HONG KONG · TAIPEI · CHENNAI · TOKYO

Published by

World Scientific Publishing Europe Ltd.
57 Shelton Street, Covent Garden, London WC2H 9HE
Head office: 5 Toh Tuck Link, Singapore 596224
USA office: 27 Warren Street, Suite 401-402, Hackensack, NJ 07601

Library of Congress Cataloging-in-Publication Data
Names: Paneva-Konovska, Jordanka.
Title: From Bessel to multi-index mittag-leffler functions : enumerable families, series in them,
    and convergence / Jordanka Paneva-Konovska (Technical University of Sofia, Bulgaria).
Description: New Jersey : World Scientific, 2016. | Includes bibliographical references and index.
Identifiers: LCCN 2016014416 | ISBN 9781786340887 (hc : alk. paper)
Subjects: LCSH: Fractional calculus. | Bessel functions.
Classification: LCC QA314 .P36 2016 | DDC 515/.55--dc23
LC record available at https://lccn.loc.gov/2016014416

British Library Cataloguing-in-Publication Data
A catalogue record for this book is available from the British Library.

Desk Editors: Kalpana Bharanikumar/Mary Simpson

Typeset by Stallion Press
Email: enquiries@stallionpress.com

Printed in Singapore

To my Family

Preface

The theory of special functions, originating from their numerous applications, is a very old branch of analysis. The long existing interest in them is recently growing in view of their new applications and further generalizations. The contemporary intensive development of this theory touches various unexpected areas of applications and is based on the tools of numerical analysis and computer algebra systems, used for analytical evaluations and graphical representations of the special functions.

However, still a greater number of mathematicians and representatives of natural, technical and others applied sciences ... 'are totally unaware of the power of special functions. They react to a paper which contains Bessel functions or Legendre polynomials by turning immediately to the next paper ... Hopefully these lectures will show how useful they could be ... So my advice is to learn something about the special functions or if this seems too hard or dull a task, get to know someone who knows something about them ... Every large university or research laboratory should have a person who not only can find things in the Bateman–Erdélyi project, but can fill in a few holes in this set of books. In any case, I hope my point has been made: special functions are useful and those who need them and those who know them should start to talk to each other.' [Askey (1975)].

The Bessel and Mittag-Leffler functions, because of their wide use and applications, have various generalizations. Among them are, for example, the 2-, 3- and 4-index analogs of the Bessel functions, hyper-Bessel functions, 3-index analog of Mittag-Leffler functions as well as the multi-index Mittag-Leffler functions.

In this book, the topics covered include enumerable families of Bessel functions, studying their completeness in the space of complex functions holomorphic in a given subset of the complex plane. Further, families of the special functions, listed above, are specified, some of their properties are established and the convergence of series in them is investigated. Results, analogical to the classical ones, for the power series are obtained, and the conclusion is that each of the considered series behaves like a power series.

The present book consists of the introduction, nine chapters and bibliography.

The introduction gives a brief historical overview of the subject matter of this book. It traces the origination of the special functions, listed above, as well as the problems concerning them, providing a motivation for further studies and applications.

In Chapter 1, we include some preliminary results on the Bessel functions and associated with them the Neumann polynomials, as definitions, integral representations and asymptotic formulae. They serve to make the book self-contained.

In Chapter 2, some results on the generating functions of the Bessel and associated Bessel functions that are necessary for use in Chapter 5 are proposed.

Chapter 3 considers the series in Bessel functions in the complex plane and studies their convergence and overconvergence, giving results analogical to the classical ones for the widely used power series, namely Cauchy–Hadamard, Abel, Tauberian, Fatou and Hadamard type theorems.

In Chapter 4, we investigate the possibility to expand a function, holomorphic in a ring, in a series in systems of generalized Neumann polynomials and Bessel functions. As a result, an analog of the Laurent's theorem is given. Some propositions related to singularities and analytical continuation of holomorphic functions, in a sense of

Pringsheim's type theorem and Vivanti–Dienes type theorem, are also established.

Chapter 5 is devoted to the completeness of different systems of Bessel and associated Bessel functions in spaces of holomorphic functions. Some auxiliary statements for their generating functions from Chapter 2, that are essentially used for proving the theorems about completeness of the corresponding system of functions, are given.

Chapters 6 and 7 deal with the multi-index generalizations of the Bessel functions of the first kind, including hyper-Bessel functions, as well as with the Mittag-Leffler functions and their three-parametric generalizations, introduced by Prabhakar. Some properties concerning the parameters of these functions, useful inequalities and some asymptotic formulae for 'large' values of indices are established for these special functions.

In Chapter 8, we study the latest generalizations of both Bessel and Mittag-Leffler type functions known in the literature as multi-index Mittag-Leffler functions. First, the $2m$-index variant is considered and the corresponding inequalities and asymptotic formulae for 'large values' of the parameters are given. Further, a definition and basic properties of the 3m-parametric Mittag-Leffler functions are given. Their representations in terms of the Fox's and Wright's functions are proved and fractional order integrals and derivatives, as the Riemann–Liuoville and generalized fractional Erdélyi–Kober, are evaluated. For illustrations, various special cases of the multi-index Mittag-Leffler functions are described.

Chapter 9 is devoted to studying the series in various enumerable systems of functions of the type, considered heretofore. In order to obtain the simplest possible results, we specify suitable countable systems of functions, by multiplying the considered functions with suitable coefficients and power functions. The main purpose is to consider the series in such families of functions in the complex plane, their disks of convergence and behavior on the boundaries, by providing statements analogous to the classical Cauchy–Hadamard, Abel, Tauber, Littlewood and Fatou type theorems, as well as overconvergence theorems, for the power series.

The Bibliography consists of 134 titles, published up to 2015. However, it does not pretend to be considered as a complete list and interested readers may find additional references in the monographs and surveys mentioned in the Introduction.

Acknowledgments

There are a number of people who have tremendously helped me with their support.

First of all, I would like to express my gratitude to my Ph.D. Scientific Advisor, Professor Peter Rusev (Institute of Mathematics and Informatics, Bulgarian Academy of Sciences, Sofia, Bulgaria), who introduced me to the theory of the Bessel functions more than a quarter century ago.

I am deeply thankful to Professor Virginia Kiryakova (Institute of Mathematics and Informatics, Bulgarian Academy of Sciences, Sofia, Bulgaria) who provoked me to write the manuscript of this book and encouraged me to submit it.

I would also like to thank the anonymous referees for their useful suggestions and recommendations in improving the manuscript. I am very obliged to the staff at the Imperial College Press and World Scientific Publishing, especially to Laurent Chaminade and Mary Simpson, for taking care of the procedures for preparation of publication of this book.

This manuscript has been prepared with the help of LaTeX typesetting system, so obviously I am extremely obliged to Donald Knuth for inventing TeX and to Leslie Lamport for creating LaTeX.

xii   *Acknowledgments*

I am also thankful to my husband Petar Konovski and to our daughter Meg for their everyday understanding, love and support.

Jordanka Paneva-Konovska

FAMI — Technical University of Sofia, Bulgaria

&IMI — Bulgarian Academy of Sciences, Bulgaria

Introduction

Bessel functions appeared in solving concrete practical problems of mechanics and astronomy. They had approved themselves as one of the most frequently used special functions in mathematical analysis, with their wide applications in physics, mechanics and engineering sciences.

They represent solutions to many problems in the above-mentioned subjects and provide an opportunity for studying them in detail, based on the broad and comprehensive information for the Bessel functions, gathered over almost two centuries.

As examples of such problems can be mentioned the axial-symmetric problems, whose analytic treating leads to linear partial differential equations (PDEs), containing the Laplace operator. Their solutions, obtained by the Fourier method of separation of the variables, in cylindric coordinates, are expressed by Bessel functions. As a result, Bessel functions, as solutions of the Bessel differential equation, are also called as cylindric functions.

In the frames of the classical complex analysis and ordinary differential equations (ODEs) theory, the Bessel functions are considered as solutions of a linear differential equation of second-order with a regular singularity at the origin and irregular singularity at the infinity. This makes the analytic nature of the Bessel functions excessively clear, which, in general, are multi-valued analytic functions with 'power' and/or 'logarithmic' singularity at the origin, respectively at the infinity.

Considering the Bessel functions as a special class of analytical functions reveals their nature. In particular, it refers to their asymptotic expansions, as well as various integral representations. As a result of numerous long-time investigations, a number of results have led to findings which are widely used in mathematical analysis and its applications.

As it is well known, studying the properties of the complex-valued functions, which are holomorphic in a complex domain, is often based on the possibility of their representations by series in concrete countable systems of functions, holomorphic in the considered domain. Taylor's systems are the most popular ones for circular domains. Their use results in the power series representations. Dirichlet type series are the ones used in a 'right' or a 'left' plane. Series in Faber polynomials are engaged in domains of more general nature. Series in classical orthogonal polynomials are also used.

One of the most frequently considered problems in complex analysis is the one for the singular points and analytic continuation of holomorphic functions, defined by the series in the respective (countable) system of such functions. This theory is most comprehensively developed for the Taylor's systems, i.e. for the power series. Together with the latter, the series in Faber polynomials, systems of exponential functions, as well as Jacobi, Laguerre and Hermite polynomials are considered. They are still an object of study.

The study of the series in the system $\{J_n(z)\}_{n=0}^{\infty}$ of Bessel functions of the first kind with nonnegative integer indices can be traced back to the 19th century works by Carl Neumann. He presented the Cauchy kernel according to bilinear series in these functions and the Neumann 'polynomials'. Using the above-mentioned series, it has been proved that the system $\{J_n(z)\}_{n=0}^{\infty}$ is a basis in the space of complex functions, holomorphic in the open disk $D(0; R)$ $(0 < R \leq \infty)$, centered at the origin and with a radius R.

The basicity problem of countable systems of Bessel functions has been proved later by other authors, for example Schäfke [1960, 1961, 1963], Lehua [1996] and others.

As it seems, the problem for the singular points and analytic continuation of holomorphic functions, defined by the series in enumerable systems of Bessel functions, has remained unnoticed. Namely, this circumstance motivates the considerations in this direction. The research is based on the fact that the system of the kind $\{z^{-\nu}J_{n+\nu}(z)\}_{n=0}^{\infty}$ ($\nu \in \mathbb{C}$) is a basis in the space of complex functions, holomorphic in the circular domain $D(0; R)$.

Further, the holomorphic functions f, g and $\tilde{g}$, defined respectively by the series in powers of z, Bessel and the modified Bessel functions of the first kind, which have the same radius R of convergence, are considered. The analytic continuation of the functions f, g and $\tilde{g}$ outside their disks of convergence is studied. Also, some relations between their main stars are found.

Pringsheim's and Vivanti–Dines type theorem are proved for the series in Bessel functions of the first kind and modified Bessel functions of the first kind. It is established that under conditions like the ones of power series, the points $z_0 = R$ and $z_0 = iR$ are singular points for $\tilde{g}(z)$ and $g(z)$ respectively.

Fundamental role in the complex analysis is played by the classical Runge's theorem (also known as Runge's approximation theorem). Introducing some denotations, we are going to recall this theorem.

Let G be a domain in the extended complex plane $\overline{\mathbb{C}}$ with a nonempty complement $G^* = \overline{\mathbb{C}}\backslash G$. Let $M \subset G^*$ be a set whose intersection with each component of G^* is a nonempty set and $R(M)$ be the set of rational fractions of the kind $(z-a)^{-m}$ ($m = 1, 2, \dots$), with $a \in M\backslash\{\infty\}$, and monomials z^n ($n = 0, 1, 2, \dots$).

The Runge's theorem states that the set $R(M)$ is complete in the space $H(G)$ of complex functions, holomorphic in the domain G, i.e. each function $f \in H(G)$ can be uniformly approximated on the compact subsets of the domain G by linear combinations of functions of $R(M)$.

In particular cases (for example, if the domain $G \subset \overline{\mathbb{C}}$ is simply connected), the countable system $\{z^n\}_{n=0}^{\infty}$ is complete in the space $H(G)$.

Numerous ongoing studies are devoted to the completeness of systems of holomorphic functions. This motivates the inclusion of results referring to the completeness of countable systems of Bessel functions.

It is proved that the above-mentioned countable system of functions $\{z^{-\nu}J_{n+\nu}(z)\}_{n=0}^{\infty}$ ($\nu \in \mathbb{C}$), as well as its subsystems, are complete in the space $H(G)$ of holomorphic functions in the simply connected domains of the complex plane. The proofs are based on the basicity of the system $\{z^{-\nu}J_{n+\nu}(z)\}_{n=0}^{\infty}$ and they use a suitable generating function of the described system. It is also established that the systems $\{K_{n+1/2}(z)\}_{n=0}^{\infty}$ of the modified Bessel functions of the third kind and $\{O_n(z)\}_{n=0}^{\infty}$ of the Neumann polynomials, respectively, as well as some of their subsystems are complete in the spaces of holomorphic functions in the respective simply connected subdomains of the complex plane $\mathbb{C}$.

Nowadays, Bessel functions have various useful generalizations obtained by adding additional indices including 2-, 3- and 4-index Bessel–Wright functions.

Between 1899 and 1905, the great Swedish mathematician Gösta Magnus Mittag-Leffler published a series of papers, which he called 'Notes', on the summation of divergent series. The aim of these 'Notes' was to construct an analytical continuation of a power series outside its disk of convergence. As it is well known, the region where he was able to do this is called Mittag-Leffler's star. In his studies, the classical Mittag-Leffler functions appear. Unfortunately, Mittag-Leffler functions remain practically unknown and unused for a long time, almost for half a century.

Although these special functions have not been widely known and used earlier, nowadays their theory can be found in the contemporary monographs [Podlubny (1999); Kilbas *et al.* (2006); Mathai and Haubold (2008); Mainardi (2010); see also Kiryakova (2000, 2008, 2010a, 2010b)]. This is because the interest in Mittag-Leffler functions has grown up together with their applications in various fractional order models which describe the phenomena of real fractal nature world more adequately than the integer order models (see e.g. [Nigmatullin and Baleanu (2012)]), and they appear as a rule, in the explicit solutions of fractional order differential and integral equations. A few such works are by Gorenflo and Mainardi [2000], Mainardi [2010], Kiryakova [2011], Baleanu *et al.* [2012], Gorenflo *et al.* [2014] and many others. Mittag-Leffler functions are also used in the statistical distributions and conditional expectations, then usually called Mittag-Leffler statistical density, see for example [Mathai and Haubold (2011)]. Due to increasing interest in the above areas and their wide and varied use, numerous generalizations of Mittag-Leffler's functions have appeared. Among them are the 2-index Mittag-Leffler functions as well as the 3-index generalized Mittag-Leffler functions, introduced by Prabhakar. Multi-index Mittag-Leffler functions (i.e. multi-index analogs of the 2-index Mittag-Leffler functions, obtained by replacing the parameters with vectors of parameters) are more general. The class of these multi-indices ($2m$-index) Mittag-Leffler functions, introduced by Yacubovich and Luchko and primarily studied in detail by Kiryakova, came into existence at the end of the 20th century. This class also includes Bessel functions and their above-listed generalizations, and Mittag-Leffler functions as well. Recently, the class of multi-index ($3m$-index) Mittag-Leffler functions has been introduced encompassing both the 3-index generalized Mittag-Leffler and $2m$-index Mittag-Leffler functions. Thus, a rich stock of generalizations including Bessel and Mittag-Leffler type functions has been produced. Further, the basic properties of these functions, such as order and type, differentiations and integrations, as well as integral representations are established.

Besides Bessel functions systems and the results referring to the convergent series in them, suitable enumerable families of the Bessel and Mittag-Leffler type functions are also specified and the series in these functions in the complex plane $\mathbb{C}$ were considered. Such results are inspired by the fact that the solutions of some fractional order differential and integral equations can be written in terms of series (or integrals, or series of integrals) of Mittag-Leffler functions, Prabhakar functions, and their multi-index generalizations, see, for example, the works [Kiryakova (2011); Sandev *et al.* (2011); Herzallah and Baleanu (2012)]. In studying their convergence, upper estimates and asymptotic formulae have been obtained, referring to the 'large' values of indices. They are further used for determining the domains of convergence, i.e. where the series converges and where it diverges, and where the convergence is uniform and where it is not. The boundary behavior of the series is also investigated, proving theorems analogical to the ones for the power series, namely the classical Abel, Tauberian and Fatou theorems, as well as the overconvergence theorems.

Regarding the results related to the convergence of the considered series in various families, obtained in this monograph, we can briefly summarize that they are completely analogical to the ones connected to the widely used power series.

Contents

Chapter 1

Bessel and Associated Functions

1.1. Bessel Functions

Solving a set of problems in mechanics and mathematical physics is closely related to the Bessel differential equation

$$z^2 \frac{d^2 w}{dz^2} + z \frac{dw}{dz} + (z^2 - \nu^2)w = 0. \tag{1.1}$$

The function $J_\nu(z)$, defined by the equality

$$J_\nu(z) = \sum_{k=0}^{\infty} \frac{(-1)^k \left(\frac{z}{2}\right)^{2k+\nu}}{k! \, \Gamma(k + \nu + 1)}, \quad z \in \mathbb{C} \backslash (-\infty, 0], \tag{1.2}$$

is a solution of this equation in the domain $\mathbb{C} \backslash (-\infty, 0]$ (see e.g. [Watson (1949, 3.1 (8))]). It is called *Bessel function of the first kind* with an index ν. The function $J_{-\nu}(z)$ is also a solution for the above equation. The Bessel functions of the first kind with an integer index are also called *Bessel coefficients*. They are holomorphic in the whole complex plane.

First, let us consider the case when the parameter ν is not integer. Then the linear combinations (see [Erdélyi *et al.* (1953, 7.2 (4)–(6))]):

$$Y_\nu(z) = \frac{J_\nu(z) \cos(\nu\pi) - J_{-\nu}(z)}{\sin(\nu\pi)}, \quad z \in \mathbb{C} \backslash (-\infty, 0], \tag{1.3}$$

$$H_\nu^{(1)}(z) = J_\nu(z) + i \, Y_\nu(z), \quad z \in \mathbb{C} \backslash (-\infty, 0], \tag{1.4}$$

$$H_\nu^{(2)}(z) = J_\nu(z) - i \, Y_\nu(z), \quad z \in \mathbb{C} \backslash (-\infty, 0] \tag{1.5}$$

1

are also solutions of the differential equation (1.1). $Y_\nu(z)$ are called *Bessel functions of the second kind*, and $H_\nu^{(1)}(z)$ and $H_\nu^{(2)}(z)$ are called *Bessel functions of the third kind*, also known as *first and second Hankel functions*.

If ν is an integer, then the left-hand side of the equalities (1.3)–(1.5) are not defined. But their limits, when $\nu \to n$ (n is an integer), exist and they can be used to define the Bessel functions of the second and third kinds with an integer index. In particular, we have $Y_n(z) = \lim_{\nu \to n} Y_\nu(z)$ (see [Erdélyi *et al.* (1953, 7.2 (28))]), i.e.

$$Y_n(z) = \frac{1}{\pi} \left[\frac{\partial J_\nu(z)}{\partial \nu} - (-1)^n \frac{\partial J_{-\nu}(z)}{\partial \nu} \right]_{\nu=n}, \quad z \in \mathbb{C}\backslash(-\infty, 0].$$

(1.6)

The functions $J_\nu(z)$ and $J_{-\nu}(z)$ form a fundamental system of solutions of the differential equation (1.1) iff ν is not an integer (see [Watson (1949, 3.12)]), whereas $J_\nu(z)$ and $Y_\nu(z)$ always form a fundamental system of solutions of this equation [Watson (1949, 3.63)]. One of the primary motives for introducing the functions $Y_\nu(z)$ is the necessity of a second solution that is linearly independent of $J_\nu(z)$, when $\nu = n$ is a nonnegative integer.

1.2. Modified Bessel Functions

The solving of some problems in mathematical physics is often related to the differential equation

$$z^2 \frac{d^2 w}{dz^2} + z \frac{dw}{dz} - (z^2 + \nu^2)w = 0$$

(1.7)

that differs from the Bessel equation by the coefficient of w and it can be obtained from (1.1) by replacing iz instead of z.

The system $\{J_\nu(iz), J_{-\nu}(iz)\}$ as well as $\{J_\nu(iz), Y_\nu(iz)\}$ are fundamental systems of solutions of Eq. (1.7) in the domain $z \in \mathbb{C}\backslash(-\infty, 0]$, but more frequently used are the functions [Watson (1949, 3.7 (2)); Erdélyi *et al.* (1953, 7.2 (12))]

$$I_\nu(z) = \sum_{k=0}^\infty \frac{\left(\frac{z}{2}\right)^{2k+\nu}}{k!\,\Gamma(k+\nu+1)}, \quad z \in \mathbb{C}\backslash(-\infty, 0]$$

(1.8)

and $I_{-\nu}(z)$. They are called *modified Bessel functions of the first kind*. The Bessel and modified Bessel functions of the first kind are related by simple dependence in the corresponding domains of the complex plane, namely:

$$I_\nu(z) = \exp\left(\frac{-i\nu\pi}{2}\right) J_\nu(iz), \quad -\pi < \arg z < \frac{\pi}{2},$$

$$I_\nu(z) = \exp\left(\frac{i\nu\pi}{2}\right) J_\nu(-iz), \quad -\frac{\pi}{2} < \arg z < \pi.$$

The functions [Erdélyi *et al.* (1953, 7.2 (13), (36))]

$$K_\nu(z) = \frac{\pi(I_{-\nu}(z) - I_\nu(z))}{2\sin(\nu\pi)}, \quad \nu \notin \mathbb{Z}, \ z \in \mathbb{C}\backslash(-\infty, 0],$$

$$K_n(z) = \frac{(-1)^n}{2}\left[\frac{\partial I_{-\nu}(z)}{\partial \nu} - \frac{\partial I_\nu(z)}{\partial \nu}\right]_{\nu=n}, \quad n \in \mathbb{Z}, \ z \in \mathbb{C}\backslash(-\infty, 0],$$

are also solutions of Eq. (1.7). They are called *modified Bessel function of the third kind*, although the modern definition is given by MacDonald [Erdélyi *et al.* (1953), 7.2.2].

The functions with an index of the kind $n + 1/2$ ($n = 0, \pm 1, \dots$), called *Bessel functions with a half-integer index* or also as *spherical Bessel functions*, form an interesting class of functions. They can be expressed as rational functions of $\sqrt{z}$, $\cos z$, $\sin z$, and $\exp z$. In particular, the modified Bessel functions of the third kind with half-integer indices satisfy the relations [Erdélyi *et al.* (1953), 7.2 (40), (42), 7.3 (16)]:

$$K_{n+\frac{1}{2}}(z) = \sqrt{\frac{\pi}{2z}} \exp(-z) \sum_{k=0}^{n} \frac{(2z)^{-k}\Gamma(n+k+1)}{k!\,\Gamma(n-k+1)}, \quad |\arg z| < \pi,$$

$$\tag{1.9}$$

$$K_{\frac{1}{2}}(z) = \sqrt{\frac{\pi}{2z}} \exp(-z), \quad |\arg z| < \pi. \tag{1.10}$$

1.3. Neumann Polynomials

The *Neumann 'polynomials'* $O_n(z)$, introduced by the relation [Watson (1949), 9.14 (2)]

$$O_n(z) = \frac{1}{2} \int_0^\infty \exp(-zt) \left((t + \sqrt{1+t^2})^n + (t - \sqrt{1+t^2})^n \right) dt,$$

$$(1.11)$$

for $|\arg z| < \pi/2$, are closely related to the Bessel functions of the first type. It is not hard to see that the so-defined function is a polynomial of $1/z$ with a power of $n+1$.

After integrating (1.11), the Neumann polynomials can be expressed explicitly in the following way (see [Erdélyi *et al.* (1953), 7.5 (4), (5)]):

$$O_{2n}(z) = \frac{n}{2} \sum_{k=0}^n \frac{(n+k-1)!}{(n-k)!} \left(\frac{z}{2}\right)^{-2k-1}, \quad n = 1, 2, \ldots, \quad (1.12)$$

$$O_{2n+1}(z) = \frac{n + \frac{1}{2}}{2} \sum_{k=0}^n \frac{(n+k)!}{(n-k)!} \left(\frac{z}{2}\right)^{-2k-2}, \quad n = 0, 1, 2, \ldots. \quad (1.13)$$

After corresponding algebraic calculations, the above representations can be written as follows (see [Erdélyi *et al.* (1953), 7.5 (6)]):

$$O_n(z) = n 2^{n-1} z^{-n-1} \sum_{k=0}^{\left[\frac{n}{2}\right]} \frac{(n-k-1)!}{k!} \left(\frac{z}{2}\right)^{2k}, \quad n = 1, 2, \ldots.$$

$$(1.14)$$

Note also that (see [Erdélyi *et al.* (1953), 7.5 (7)])

$$O_0(z) = \frac{1}{z}. \quad (1.15)$$

Gegenbauer generalized the Neumann polynomials by introducing the polynomials $A_{n,\nu}(z)$ (see [Watson (1949), 9.2 (1)]) as follows:

$$A_{n,\nu}(z) = 2^{n+\nu} z^{-n-1} (n+\nu) \sum_{k=0}^{\left[\frac{n}{2}\right]} \frac{\Gamma(n+\nu-k)}{k!} \left(\frac{z}{2}\right)^{2k}, \quad (1.16)$$

for $\nu \neq 0, -1, -2, \ldots$.

1.4. Integral Representations and Asymptotic Formulae

1.4.1. Preliminary results

One of the most often used representations of the Bessel functions of the first kind is the Poisson integral representation [Erdélyi *et al.* (1953), 7.12 (7)]:

$$J_\nu(z) = \frac{z^\nu}{\sqrt{\pi} 2^\nu \Gamma\left(\nu + \frac{1}{2}\right)} \int_{-1}^{1} (1 - t^2)^{\nu - \frac{1}{2}} \exp(izt) dt \qquad (1.17)$$

that holds when $\Re(\nu) > -1/2$. It serves as the origin of many important investigations in the area of Bessel functions.

The modified Bessel functions of the third kind have the representation [Erdélyi *et al.* (1953), 7.12 (27)]:

$$K_\nu(z) = \frac{(2z)^\nu \Gamma\left(\nu + \frac{1}{2}\right)}{\sqrt{\pi}} \int_{0}^{\infty} (z^2 + t^2)^{-\nu - \frac{1}{2}} \cos t \, dt \qquad (1.18)$$

that holds when $\Re(\nu) > -1/2$ and $|\arg z| < \pi/2$, i.e. when z lies in the right half-plane $\Re(z) > 0$ and $\Re(\nu) > -1/2$.

Depending on the above representation, whether the index ν or the argument z grows infinitely, different asymptotic formulae are known for the Bessel functions. So, for example, the Bessel coefficients $J_n(z)$ have the following representation (see e.g. [Whittaker and Watson (1963), §17.81]):

$$J_n(z) = \left(\frac{z}{2}\right)^n (1 + \theta_n(z)) \frac{1}{n!}, \quad \theta_n(z) \to 0 \text{ as } n \to \infty, \qquad (1.19)$$

in the whole complex plane. The functions $\theta_n(z)$ are holomorphic for $z \in \mathbb{C}$ and moreover $\lim_{n\to\infty} \theta_n(z) = 0$ uniformly on each compact subset of the plane $\mathbb{C}$.

The Bessel functions have the following representation when $z \to \infty$ [Erdélyi *et al.* (1953), 7.13 (3)]:

$$J_\nu(z) = \sqrt{\frac{2}{\pi z}} \left(\cos(z - \lambda_\nu) - \sin(z - \lambda_\nu) O\left(\frac{1}{|z|}\right) \right), \qquad (1.20)$$

with $\lambda_\nu = \frac{\nu\pi}{2} + \frac{\pi}{4}$, for $\arg|z| \le \pi - \delta$ and arbitrary $0 < \delta < \pi$.

For the Neumann polynomials, the following asymptotic formula [Whittaker and Watson (1963), §17.81] is known:

$$O_n(z) = n!2^{n-1}z^{-n-1}(1 + \varphi_n(z)), \quad \lim_{n\to\infty} \varphi_n(z) = 0, \qquad (1.21)$$

when $z \in \overline{\mathbb{C}}\backslash\{0\}$.

Considering the explicit form (1.16) of the Neumann polynomials $A_{n,\nu}(z)$, it is not hard to obtain the asymptotic formula

$$A_{n,\nu}(z) = 2^{n+\nu}z^{-n-1}\Gamma(n + \nu + 1)(1 + \phi_n(z)), \quad \lim_{n\to\infty} \phi_n(z) = 0,$$
$$(1.22)$$

valid again for $z \in \overline{\mathbb{C}}\backslash\{0\}$.

The convergence in (1.21) and (1.22) is uniform on the compact subsets of the domain $\overline{\mathbb{C}}\backslash\{0\}$.

1.4.2. An upper estimate

Considering explicitly $\theta_n(z)$, we can make the result from (1.19), sharper.

Theorem 1.1. *Let $K \subset \mathbb{C}$ be a nonempty compact set. Then there exists a constant $C = C(K)$, $0 < C < \infty$, such that for each $n \in \mathbb{N}_0$ and each $z \in K$, the following inequality holds:*

$$|\theta_n(z)| \le \frac{C}{(n+1)}. \qquad (1.23)$$

Proof. First, let $z \in \mathbb{C}$. As a result of (1.2) and (1.19), we can write

$$\theta_n(z) = \frac{1}{n+1}\sum_{k=1}^{\infty}\frac{(-1)^k(n+1)!}{k!(n+k)!}\left(\frac{z}{2}\right)^{2k}.$$

Denoting $u_k(z) = \frac{(-1)^k(n+1)!}{k!(n+k)!}\left(\frac{z}{2}\right)^{2k}$, we obtain the estimate

$$|u_k(z)| \le \frac{1}{k!}\left|\frac{z}{2}\right|^{2k} \qquad (1.24)$$

for the absolute value of $u_k(z)$. The series

$$\sum_{k=1}^{\infty}\frac{1}{k!}\left|\frac{z}{2}\right|^{2k} \qquad (1.25)$$

converges for each $z \in \mathbb{C}$ and its sum is $\exp(|z|^2/4) - 1$. This shows that

$$|\theta_n(z)| \leq \frac{1}{n+1} \left(\exp\left(\frac{|z|^2}{4}\right) - 1 \right) \tag{1.26}$$

on the whole complex plane.

Then, the estimate (1.23) follows immediately from (1.26) for all the values $z \in K$. □

The proof exposed above can also be seen in [Paneva-Konovska (2009)]. Furthermore, the following remarks can be made.

Remark 1.1. The uniform convergence of $\theta_n(z)$ on the compact subsets of $\mathbb{C}$ follows from (1.23) as well.

Remark 1.2. According to the asymptotic formula (1.19), it follows that there exists a natural number N_0 such that the functions $J_n(z)$ have no zeros for $n > N_0$, except for the origin.

Remark 1.3. Note that each of the functions $J_n(z)$ $(n \in \mathbb{N}_0)$, being an entire function, not identically zero, has no more than a finite number of zeros in the closed and bounded set $|z| \leq R$ [Markushevich (1967), Vol. 1, p. 305, Chapter 3, §6, 6.1]. Moreover, because of Remark 1.2, no more than a finite number of these functions have some zeros, except for the origin.

Remark 1.4. The formula (1.23) is essentially used in proving the corresponding strengthened version of Tauber type theorem for the series in Bessel functions of the first kind.

Chapter 2

Generating Functions of Bessel and Associated Bessel Functions

In this chapter, which is based on the paper [Paneva-Konovska (2003)], we give the generating functions for countable families of Bessel functions of the first kind, modified Bessel functions of the third kind with half-integer indices and Neumann polynomials and a few results which follow from these generating functions. We need them to prove the completeness in the spaces of holomorphic functions.

In what follows, we shall use the notations $\Phi(z, w)$ for above-mentioned functions and cite them with their labels.

2.1. Generating Functions of Bessel Functions of the First Kind

Let $J_\nu(z)$ be the Bessel functions of the first kind with an index ν, given by (1.2). Denote

$$\Phi(z, w) = \pi^{-\frac{1}{2}} \int_{-1}^{1} \frac{\exp(izt)}{1 - zw(1 - t^2)} (1 - t^2)^{\nu - \frac{1}{2}} dt,$$

$$\text{with} \quad 1 - zw(1 - t^2) \neq 0, \quad \text{and} \quad \Re(\nu) > -\frac{1}{2}. \quad (2.1)$$

Theorem 2.1. *If* $\Re(\nu) > -1/2$, $z \in \mathbb{C}$, $w \in \mathbb{C}$, $|zw| < 1$ *and* $\Phi(z,w)$ *is the function, defined by* (2.1), *then the following expansion holds:*

$$\Phi(z,w) = \sum_{n=0}^{\infty} 2^{n+\nu} \Gamma\left(n + \nu + \frac{1}{2}\right) z^{-\nu} J_{n+\nu}(z) w^n. \qquad (2.2)$$

Proof. In order to prove this theorem we use the Poisson integral representation (1.17), applied to the function $J_{n+\nu}(z)$. Then

$$\Gamma\left(n + \nu + \frac{1}{2}\right) z^{-\nu} J_{n+\nu}(z)$$

$$= \frac{z^n}{\sqrt{\pi} 2^{n+\nu}} \int_{-1}^{1} \exp(izt) \left(1 - t^2\right)^{n+\nu-\frac{1}{2}} dt, \qquad (2.3)$$

from which the right-hand side of (2.2) becomes

$$\sum_{n=0}^{\infty} 2^{n+\nu} \Gamma\left(n + \nu + \frac{1}{2}\right) z^{-\nu} J_{n+\nu}(z) w^n$$

$$= \frac{1}{\sqrt{\pi}} \sum_{n=0}^{\infty} (zw)^n \int_{-1}^{1} \exp(izt) \left(1 - t^2\right)^{n+\nu-\frac{1}{2}} dt. \qquad (2.4)$$

Since the inequality

$$\left| \exp(izt) \left(1 - t^2\right)^{n+\nu-\frac{1}{2}} (zw)^n \right| \leq \left(1 - t^2\right)^{\Re\left(\nu+\frac{1}{2}\right)} |zw|^n \exp|z|$$

holds for $t \in [-1, 1]$ and $n = 1, 2, \ldots$ and the series $\sum_{n=1}^{\infty} |zw|^n$ converges, then the series

$$\sum_{n=1}^{\infty} (zw)^n \exp(izt) \left(1 - t^2\right)^{n+\nu-\frac{1}{2}}$$

is absolutely and uniformly convergent with respect to t in the interval $[-1, 1]$. Then the change of the order of the integration and summation leads to the identities:

$$\sum_{n=0}^{\infty} (zw)^n \left(1 - t^2\right)^n = \frac{1}{1 - zw\left(1 - t^2\right)},$$

$$\sum_{n=0}^{\infty} 2^{n+\nu}\Gamma\left(n + \nu + \frac{1}{2}\right) z^{-\nu} J_{n+\nu}(z)w^n$$

$$= \frac{1}{\sqrt{\pi}} \int_{-1}^{1} \frac{\exp(izt)}{1 - zw\left(1 - t^2\right)} \left(1 - t^2\right)^{\nu - \frac{1}{2}} dt,$$

that proves (2.2). $\qquad\square$

2.2. Generating Functions of Bessel Functions of the Third Kind with Half-Integer Indices

Let $K_{n+1/2}(z)$, $n = 0, 1, 2, \ldots$, be the modified Bessel functions (1.9) of the third kind with a half-integer index.

Let $0 < \alpha < 1$, A_α be the set

$$A_\alpha = \{z : z \in \mathbb{C}, |\arg z| \leq \pi\alpha/2\} \tag{2.5}$$

and A be a subset of A_α with the condition

$$r = \inf_{z \in A} |z| > 0. \tag{2.6}$$

Define ρ_0 as follows

$$\rho_0 = \begin{cases} r, & \text{for } 0 < \alpha \leq \dfrac{1}{2}, \\[2mm] r\sin\pi\alpha, & \text{for } \dfrac{1}{2} < \alpha < 1. \end{cases} \tag{2.7}$$

Lemma 2.1. *Let $0 < \alpha < 1$, A_α be a set of the kind (2.5), the set $A \subset A_\alpha$ satisfy the condition (2.6) and ρ_0 be defined by the equality (2.7). Then the following inequality holds:*

$$\left|z + z^{-1}t^2\right| \geq \rho_0 \tag{2.8}$$

for each $z \in A$ and $t \in [0, \infty)$.

Proof. Let $z = \rho \exp(i\vartheta)$. Then $z + z^{-1}t^2 = (\rho + \rho^{-1}t^2)\cos\vartheta + i(\rho - \rho^{-1}t^2)\sin\vartheta$ and

$$\left| z + z^{-1}t^2 \right|^2 = \rho^2 + \rho^{-2}t^4 + 2t^2\cos(2\vartheta). \tag{2.9}$$

Furthermore, we are going to divide the proof into two parts depending on the values of ϑ.

(i) Let $|\vartheta| \leq \pi/4$. Therefore, $\cos(2\vartheta) \geq 0$, from which it follows that

$$\left| z + z^{-1}t^2 \right|^2 \geq \rho^2. \tag{2.10}$$

(ii) Let $\pi/4 < |\vartheta| \leq \pi/2$. Setting $\tau = t^2$, we denote the right-hand side of (2.9) by $f(\tau)$. We have $f(\tau) = \rho^{-2}\tau^2 + 2\tau\cos(2\vartheta) + \rho^2$. Since the derivative $f'(\tau) = 2\rho^{-2}\tau + 2\cos(2\vartheta)$ is equal to zero for $\tau_0 = -\rho^2\cos(2\vartheta)$, then $f(\tau)$ has its minimal value, i.e.

$$
\begin{aligned}
f_{\min} &= f(\tau_0) \\
&= \rho^{-2}\left(-\rho^2\cos(2\vartheta)\right)^2 + 2\left(-\rho^2\cos(2\vartheta)\right)\cos(2\vartheta) + \rho^2 \\
&= \rho^2 - \rho^2\cos^2(2\vartheta) = \rho^2\sin^2(2\vartheta).
\end{aligned}
$$

Therefore,

$$\left| z + z^{-1}t^2 \right|^2 \geq \rho^2\sin^2(2\vartheta). \tag{2.11}$$

Let now $z \in A$ and $0 < \alpha \leq 1/2$. Then $|\vartheta| \leq \pi/4$ and $\left| z + z^{-1}t^2 \right| \geq \rho$, according to (2.10). Therefore, in accordance with (2.6) and (2.7), it follows that (2.8) holds. If $1/2 < \alpha < 1$, from (2.10) and (2.11) we conclude respectively that $\left| z + z^{-1}t^2 \right| \geq r$ for $|\vartheta| \leq \pi/4$ and $\left| z + z^{-1}t^2 \right| \geq r\sin(\pi\alpha)$ for $\pi/4 < |\vartheta| \leq \pi/2$, from which (2.8) holds in this case, as well. □

Remark 2.1. Note that if K is a compact (nonempty) subset of the half-plane $\{z : z \in \mathbb{C}, \Re(z) > 0\}$, then

$$\inf_{z \in K} |z| = r > 0, \quad \sup_{z \in K} |z| = R < \infty.$$

Lemma 2.2. *Let K be a compact subset of A_α, defined by (2.5), $\inf_{z \in K} |z| = r$, ρ_0 be defined by the equality (2.7), $z \in K$ and $|w| < \rho_0$.*

Then the following equality holds:

$$\sum_{n=0}^{\infty} \int_0^\infty \frac{\cos t}{z^2 + t^2} \left(\frac{zw}{z^2 + t^2} \right)^n dt = \int_0^\infty \frac{\cos t}{z^2 + t^2} \sum_{n=0}^{\infty} \left(\frac{zw}{z^2 + t^2} \right)^n dt,$$

$$(2.12)$$

for every $z \in K$ and $t \in [0, \infty)$.

Proof. First, let us prove that if $0 < c < \infty$, then the next equality holds:

$$\sum_{n=0}^{\infty} \int_0^c \frac{\cos t}{z^2 + t^2} \left(\frac{zw}{z^2 + t^2} \right)^n dt = \int_0^c \frac{\cos t}{z^2 + t^2} \sum_{n=0}^{\infty} \left(\frac{zw}{z^2 + t^2} \right)^n dt.$$

$$(2.13)$$

Denote

$$u_n(z, w; t) = \frac{\cos t}{z^2 + t^2} \left(\frac{zw}{z^2 + t^2} \right)^n, \quad t \in [0, c].$$

Bearing in mind Eq. (2.8), we have

$$|u_n(z, w; t)| \leq \left| z^2 + t^2 \right|^{-1} \left| w \left(z + z^{-1} t^2 \right)^{-1} \right|^n \leq M \left(|w| \rho_0^{-1} \right)^n,$$

$$(2.14)$$

with $M = \sup \left| z^2 + t^2 \right|^{-1}$, $z \in K$ and $t \in [0, c]$. From the convergence of the series $\sum_{n=0}^{\infty} \left(|w| \rho_0^{-1} \right)^n$ and the inequality (2.14), it follows that the series $\sum_{n=0}^{\infty} u_n(z, w; t)$ is absolutely and uniformly convergent for $t \in [0, c]$. Therefore, the equality (2.13) holds.

The last thing that still remains to be proved is that the series

$$\sum_{n=0}^{\infty} \int_0^\infty \left| \frac{\cos t}{z^2 + t^2} \left(\frac{zw}{z^2 + t^2} \right)^n \right| dt \qquad (2.15)$$

converges. For this purpose, denote the general term of the series (2.15) with $v_n(z, w)$ and let $R = \sup_{z \in K} |z|$. We have

$$
\begin{aligned}
v_n(z, w) &= \int_0^\infty \left| \frac{\cos t}{z^2 + t^2} \left(\frac{zw}{z^2 + t^2} \right)^n \right| dt \\
&= \int_0^{R+1} \left| \frac{\cos t}{z^2 + t^2} \left(\frac{zw}{z^2 + t^2} \right)^n \right| dt \\
&\quad + \int_{R+1}^\infty \left| \frac{\cos t}{z^2 + t^2} \left(\frac{zw}{z^2 + t^2} \right)^n \right| dt \\
&\leq \int_0^{R+1} \left| \frac{\cos t}{z^2 + t^2} \right| \left(\frac{|w|}{\rho_0} \right)^n dt + \int_{R+1}^\infty \frac{1}{t^2 - R^2} \left(\frac{|w|}{\rho_0} \right)^n dt \\
&\leq \left(\int_0^{R+1} \frac{dt}{|z^2 + t^2|} + \int_{R+1}^\infty \frac{dt}{t^2 - R^2} \right) \left(\frac{|w|}{\rho_0} \right)^n \leq N \left(\frac{|w|}{\rho_0} \right)^n,
\end{aligned}
$$

where $N = (R + 1) \max_{K \times [0, R+1]} |z^2 + t^2|^{-1} + \frac{\ln(2R+1)}{2R}$. The last inequality shows that the series (2.15) converges. Because of this and (2.13), it follows that (2.12) holds. $\qquad\square$

Introducing the denotation

$$
\Phi(z, w) = \sqrt{\frac{z}{\pi}} \int_0^\infty \frac{\cos t}{z^2 + t^2 - zw} \, dt, \quad w \neq z + z^{-1} t^2, \ t \in [0, \infty),
\tag{2.16}
$$

we use it in the next few statements.

Theorem 2.2. *Let K be a compact subset of A_α, defined by (2.5), $\inf_{z \in K} |z| = r$, ρ_0 be defined by the equality (2.7), $z \in K$, $|w| < \rho_0$ and $\Phi(z, w)$ be the function (2.16).*

Then the following equality holds:

$$
\Phi(z, w) = \sum_{n=0}^\infty \frac{K_{n+\frac{1}{2}}(z) \, w^n}{2^{n+\frac{1}{2}} \, \Gamma(n + 1)},
\tag{2.17}
$$

for every $z \in K$ and $t \in [0, \infty)$.

Proof. Because of (1.18), applied for $\nu = n + 1/2$, $n = 0, 1, 2, \ldots$, we obtain consecutively

$$K_{n+\frac{1}{2}}(z) = \pi^{-\frac{1}{2}}(2z)^{n+\frac{1}{2}}\Gamma(n+1) \int_0^\infty (z^2 + t^2)^{-n-1} \cos t\, dt,$$

$$\sum_{n=0}^\infty \frac{K_{n+\frac{1}{2}}(z)\, w^n}{2^{n+\frac{1}{2}}\, \Gamma(n+1)} = \sqrt{\frac{z}{\pi}} \sum_{n=0}^\infty \int_0^\infty \frac{\cos t}{z^2 + t^2} \left(\frac{zw}{z^2 + t^2} \right)^n dt,$$

and from (2.12) it follows that

$$\sum_{n=0}^\infty \frac{K_{n+\frac{1}{2}}(z)\, w^n}{2^{n+\frac{1}{2}}\, \Gamma(n+1)} = \sqrt{\frac{z}{\pi}} \int_0^\infty \sum_{n=0}^\infty \frac{\cos t}{z^2 + t^2} \left(\frac{zw}{z^2 + t^2} \right)^n dt.$$

From (2.8) follows that for every $t \in [0, \infty)$

$$\left| zw \left(z^2 + t^2 \right)^{-1} \right| = \left| w \left(z + z^{-1}t^2 \right)^{-1} \right| \le |w|\rho_0^{-1} < 1, \qquad (2.18)$$

that leads to the equality

$$\sum_{n=0}^\infty \left(\frac{zw}{z^2 + t^2} \right)^n = \frac{1}{1 - zw(z^2 + t^2)^{-1}}.$$

Then

$$\sum_{n=0}^\infty \frac{K_{n+\frac{1}{2}}(z)\, w^n}{2^{n+\frac{1}{2}}\, \Gamma(n+1)} = \sqrt{\frac{z}{\pi}} \int_0^\infty \frac{\cos t}{z^2 + t^2} \frac{1}{1 - zw(z^2 + t^2)^{-1}}\, dt,$$

from which Eq. (2.17) follows immediately.

Note that from (2.18) follows $w \ne z + z^{-1}t^2$, $t \in [0, \infty)$. $\qquad\square$

Theorem 2.3. *Let $0 < \alpha < 1$, K be a compact subset of A_α, defined by (2.5), $\inf_{z \in K} |z| = r$, ρ_0 be defined by the equality (2.7), $z \in K$, $|w| < \rho_0$ and $\Phi(z, w)$ be the function, defined by (2.16).*

Then the following equality holds:

$$\Phi(z, w) = \frac{1}{2}\sqrt{\frac{\pi}{z - w}} \exp(-\sqrt{z}\sqrt{z - w}), \qquad (2.19)$$

for every $z \in K$ and $t \in [0, \infty)$.

Proof. Applying (1.18) for $\nu = 1/2$, we have

$$K_{\frac{1}{2}}(z) = \sqrt{\frac{2z}{\pi}} \int_0^\infty (z^2 + t^2)^{-1} \cos t \, dt, \quad \Re(z) > 0.$$

Subsequently, replacing $\zeta = \sqrt{z}\sqrt{z - w}$, we reduce the integral in (2.16) to the form:

$$\int_0^\infty \frac{\cos t}{z^2 + t^2 - zw} \, dt = \int_0^\infty \left(\zeta^2 + t^2\right)^{-1} \cos t \, dt$$

$$= \sqrt{\frac{\pi}{2\zeta}} K_{\frac{1}{2}}(\zeta), \quad \Re(\zeta) > 0,$$

and expressing $K_{1/2}(\zeta)$ from (1.10), we get to the equality (2.19) (for $\Re(\zeta) > 0$).

Now, we need proof that the relation (2.19) holds for all the values $z \in K$ and $|w| < \rho_0$. For this purpose, we are going to start with the holomorphicity of the function on the right-hand side of equality (2.19). So, letting $z = x + iy$ and $w = u + iv$, without loss of generality, we can consider $y = v$, from which $z - w = x - u + i(y - v) = x - u$. Further, if $u \leq 0$ then $x - u > 0$. If $u > 0$, since $|z| > |w|$, again $x - u > 0$. Therefore, $z - w \in \mathbb{C} \setminus (-\infty, 0]$, from which the right-hand side of (2.19) is a holomorphic function whenever $|w| < \rho_0$.

Let now $z = x + iy$, $w = u < 0$, then $x - u > 0$ and $|\arg z| < \pi/2$, $|\arg(z - w)| < \pi/2$ from which $|\arg \zeta| = |\arg \sqrt{z} + \arg \sqrt{z - w}| < \pi/4 + \pi/4 = \pi/2$, i.e. $\Re(\zeta) > 0$. Therefore, in this case the right-hand sides of (2.16) and (2.19) coincide.

Hence, from the theorem for identity of two holomorphic functions, it follows that the right-hand sides of (2.16) and (2.19) coincide for every $z \in K$ and $|w| < \rho_0$ that proves Theorem 2.3. $\square$

2.3. Generating Functions of the Neumann Polynomials

Denote

$$\Phi(z, w) = \frac{1}{2} \int_0^\infty \exp(-zt)$$

$$\times \left(\exp(-w(t + \sqrt{1 + t^2})) + \exp(-w(t - \sqrt{1 + t^2})) \right) dt. \tag{2.20}$$

Theorem 2.4. *Let*

$$z, w \in \mathbb{C} : \Re(z) > 0, \ \Re(z + 2w) > 0. \tag{2.21}$$

Then $\Phi(z, w)$ is a holomorphic function of the variables z and w in the set (2.21).

Proof. Denoting $x = \Re(z)$ and $u = \Re(w)$, we can prove that the integral on the right-hand side of (2.20) is a sum of two integrals, which are absolutely and uniformly convergent on the compact subsets of the region (2.21), and therefore it is absolutely and uniformly convergent there, as well. To this purpose, denoting

$$I_1(x, u) = \int_0^\infty \exp(-t(x + 2u)) \exp\left(u(t - \sqrt{1 + t^2})\right) dt,$$

$$I_2(x, u) = \int_0^\infty \exp(-tx) \exp\left(u(\sqrt{1 + t^2} - t)\right) dt,$$

we have

$$I_1(x, u) \leq \int_0^\infty \exp(-t(x + 2u)) \exp\left(|u|(\sqrt{1 + t^2} - t)\right) dt,$$

$$I_2(x, u) \leq \int_0^\infty \exp(-tx) \exp\left(|u|(\sqrt{1 + t^2} - t)\right) dt.$$

Since the function $\exp\left(|u|(\sqrt{1+t^2} - t)\right)$ is decreasing for $t \in [0, \infty)$ and $|u|(\sqrt{1+t^2} - t))|_{t=0} = |u|$, then $\exp\left(|u|(\sqrt{1+t^2} - t)\right) \leq \exp|u|$, from which we obtain respectively

$$I_1(x, u) \leq \int_0^\infty \exp(-t(x + 2u)) \exp|u|\, dt, \qquad (2.22)$$

$$I_2(x, u) \leq \int_0^\infty \exp(-tx) \exp|u|\, dt, \qquad (2.23)$$

$$2|\Phi(z, w)|$$
$$= \left| \int_0^\infty \exp(-zt) \left(\exp(-w(t + \sqrt{1+t^2})) \right. \right.$$
$$\left. \left. + \exp(-w(t - \sqrt{1+t^2})) \right) dt \right| \qquad (2.24)$$
$$\leq I_1(x, u) + I_2(x, u).$$

Furthermore, let K be a compact subset of the domain, defined by (2.21), and (z, w) be an arbitrary point of K. Then there exist positive numbers δ, ε and M such that the inequalities $x + 2u > \delta$, $x > \varepsilon$ and $\exp|u| \leq M$ hold, provided $(z, w) \in K$. Since $\exp(-t(x + 2u)) \leq \exp(-t\delta)$, $\exp(-tx) \leq \exp(-t\varepsilon)$ and

$$\int_0^\infty \exp(-t\delta) = \frac{1}{\delta}, \quad \int_0^\infty \exp(-t\varepsilon) = \frac{1}{\varepsilon},$$

then the integrals $I_1(x, u)$ and $I_2(x, u)$ are uniformly convergent in the compact set K.

Keeping in view the inequalities (2.22)–(2.24), we can conclude that the function $\Phi(z, w)$, defined by (2.20), is holomorphic with respect to its two variables in the domain (2.21). $\qquad \square$

Finally, note that the function (2.20), by performing n times partial differentiation with respect to the variable w (calculated at the point $w = 0$), generates the Neumann polynomials $O_n(z)$.

Chapter 3

Convergence of Series in Bessel Functions

3.1. Some Sets in the Complex Plane, Denotations and One Useful Geometric Inequality

In this section, we are going to describe several kinds of sets in the complex plane, used in the following.

Beginning with the simplest circular domains, we denote with $D(0; R)$ the open disk, centered at the origin and with a radius R, $[D(0; R)]$ its closure and $C(0; R)$ its boundary, i.e.

$$D(0; R) = \{z : z \in \mathbb{C}, |z| < R\},$$
$$[D(0; R)] = D(0; R) \cup C(0; R), \tag{3.1}$$
$$C(0; R) = \partial D(0; R) = \{z : z \in \mathbb{C}, |z| = R\}.$$

Actually, the set $[D(0; R)]$ is the closed disk $\{z : z \in \mathbb{C}, |z| \leq R\}$.

Furthermore, letting $0 \leq \psi < 2\pi$ and $0 < \varphi < \pi/2$, denote with $A_{\psi,\varphi}$ the angular domain, given as follows:

$$A_{\psi,\varphi} = \{z : z \in \mathbb{C}, |\arg(z \exp(-i\psi))| < \varphi\}. \tag{3.2}$$

This domain has the size of an angle $2\varphi < \pi$ and it is symmetric with respect to the ray, passing through the origin and having an angle ψ with the positive real axis. If $0 \leq \psi < \pi$, the sets $A_{\psi,\varphi}$ and $A_{\psi+\pi,\varphi}$ form a pair of vertically opposite angles, i.e. both angles share a vertex (the origin) and their arms are symmetric with respect to the straight line, passing through the origin and having an angle ψ with the positive real axis. For example, the angles $A_{0,\varphi}$ and $A_{\pi,\varphi}$

are vertically opposite angles with the vertices at the origin and each of them is symmetric with respect to the real axis, i.e. the real axis is a bisectrix of each of them.

In order to consider the next set, taking the complex number $z_0 \neq 0$ with $\arg z_0 = \psi$, we denote

$$
\begin{aligned}
A_{z_0;\,\psi,\varphi} &= \{z : z \in \mathbb{C},\ z - z_0 \in A_{\psi+\pi,\varphi}\} \quad \text{if } 0 \leq \psi < \pi, \\
A_{z_0;\,\psi,\varphi} &= \{z : z \in \mathbb{C},\ z - z_0 \in A_{\psi-\pi,\varphi}\} \quad \text{if } \pi \leq \psi < 2\pi.
\end{aligned}
\tag{3.3}
$$

In fact, $A_{z_0;\,\psi,\varphi}$ is obtained as an image of the domain $A_{\psi\pm\pi,\varphi}$ with a translation, i.e. pictured the origin at the point z_0. It is an angular domain, containing the origin, with a size of 2φ and with a vertex at the point $z = z_0$ that is symmetric with respect to the straight line passing through the points 0 and z_0.

Finally, the denotations g_φ, $D(0; R_t)$ and d_φ are used respectively for the part (see Fig. 3.1) of the angular domain $A_{z_0;\,\psi,\varphi}$ that is included in the open disk $D(0; R)$, disk centered at the origin and touching the arms of the angle, and part of the domain g_φ, closed between the angle arms and the arc of the circle $C(0; R_t)$ (see Fig. 3.2), i.e.

$$
\begin{aligned}
g_\varphi &= g_{z_0;\varphi} = D(0; R) \cap A_{z_0;\,\psi,\varphi}, \\
d_\varphi &= d_{z_0;\varphi} = \mathrm{Ext}D(0; R_t) \cap g_\varphi.
\end{aligned}
\tag{3.4}
$$

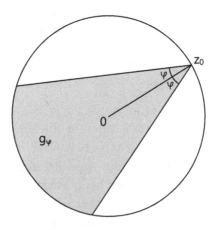

Fig. 3.1. The domain g_φ.

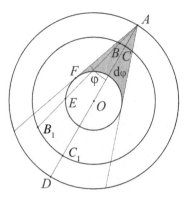

Fig. 3.2. The domain d_φ.

Let $z_0 \in \mathbb{C}$, $0 < R < \infty$, $|z_0| = R$, g_φ be an arbitrary angle domain like (3.4), with a size of $2\varphi < \pi$ and with a vertex at the point $z = z_0$, and d_φ be the corresponding set defined by (3.4).

A useful geometric inequality concerning the set d_φ is given below.

Lemma 3.1. *Let d_φ be the part of the angular domain g_φ, defined by (3.4), and $z \in [d_\varphi]$. Then*

$$|z - z_0| \cos \varphi < 2(|z_0| - |z|). \tag{3.5}$$

Proof. First, let z belong to the open set d_φ. Picturing z_0 and z respectively with the points A and B (see Fig. 3.2), we can write $AB \cdot AB_1 = AC \cdot AC_1$ from which

$$\frac{AB}{AC} = \frac{|z - z_0|}{|z_0| - |z|} = \frac{AC_1}{AB_1} < \frac{AD}{AE} < \frac{AD}{AF} = \frac{2AO}{AF} = \frac{2}{\cos \varphi}.$$

This proves the lemma in the case $z \in d_\varphi$. When $z \in \partial d_\varphi$, the proof goes in the same way, using similar inequalities.   □

3.2. Classical Results for the Power Series

An important property of the holomorphic functions is their potential to be expanded by a power series

$$\sum_{n=0}^{\infty} a_n z^n, \quad a_n \in \mathbb{C}, \ n = 0, 1, 2, \ldots. \tag{3.6}$$

Some useful information on the convergence of such a type of series in complex domains is given by the classical Cauchy–Hadamard, Abel, Tauber, and Fatou theorems.

So, according to the Cauchy–Hadamard theorem, each series of the kind (3.6) is absolutely convergent in the disk $D(0; R)$ with a radius $R = (\limsup_{n\to\infty} |a_n|^{1/n})^{-1}$ and divergent on its outside, $|z| > R$. In general, according to Abel's theorem, from the convergence of a power series $f(z) = \sum_{n=0}^{\infty} a_n z^n$ at a point z_0, there follows the existence of the limit $\lim_{z\to z_0} f(z) = f(z_0)$, when z belongs to a suitable angle domain with a vertex at a point z_0. The geometrical series [Tchakalov (1972), p. 92]: $\frac{1}{1+z} = 1 - z + z^2 - z^3 + \cdots$ at $z_0 = 1$ gives an example that in general, the inverse proposition is not true, i.e. the existence of the limit, mentioned above, does not imply the convergence of the series $\sum_{n=0}^{\infty} a_n z_0^n$ without additional conditions on the growth of the coefficients.

The origin of the summability theory of divergent series can be traced back to the time of Leonard Euler. He was the first to discuss the question of 'how-to define the sum of divergent series', rather than 'what' equals this sum. He wrote ([Euler (1787, Chapter III, pages 78, 79, points 110, 111)]): 'I say that the whole difficulty lies in the sense of the word "sum". So, if the phrase "sum of series" means, as usual, the result of the summation of all its terms, then a sum can be obtained only for convergent series. If the word "sum" is understood only in such a close sense, then the divergent series have not any sum, at all. If another sense, different from the usual, is given to the word "sum", one can get out of these difficulties. Moreover, if the series is convergent, this "new" sum is required to be the same as the usual one. Since the divergent series do not have any sum, the new sum makes no problems'. Euler's point of view for divergent series is completely actual: A divergent series has no sum in the usual sense, but it is possible to introduce a new definition of the 'sum' (i.e. definition of the summation method of series), applied to all the convergent and some divergent series. Moreover, this requires the new sum of the convergent series to be the same as in an usual sense (i.e. the method to be regular).

Let us consider the numerical series

$$\sum_{n=0}^{\infty} a_n, \quad a_n \in \mathbb{C}, \ n = 0, 1, 2, \ldots. \tag{3.7}$$

To define its Abel summability [Hardy (1949, p. 20, 1.3 (2))], we also attract the power series (3.6).

Definition 3.1. The series (3.7) is called *A-summable*, if the power series (3.6) converges in the unit disk $D(0; 1)$ and moreover there exists the limit

$$\lim_{z \to 1-0} \sum_{n=0}^{\infty} a_n z^n = S.$$

The complex number S is called *A-sum* of the considered numerical series (3.7) and the usual notation of that is

$$\sum_{n=0}^{\infty} a_n = S \qquad (A).$$

Clearly, this definition refers to both convergent and some divergent series. But according to Abel's theorem, the Abel summability is regular, i.e. the A-sum of each convergent numerical series is equal to its standard sum. So, the following remarks can be made.

Remark 3.1. The A-summation is regular. It means that if the series (3.7) is convergent, then it is A-summable, and its A-sum is equal to its usual sum.

Remark 3.2. The A-summability of the series (3.7) does not imply in general its convergence. But, with additional conditions on the growth of the general term of the series (3.7), the convergence can be ensured.

Now, we are formulating the result, reverse of the Abel's theorem, as it is given in Hardy [1949, Theorem 85].

Theorem 3.1 (of Tauber). *Let the numerical series* (3.7) *be A-summable,*

$$\sum_{n=0}^{\infty} a_n = S \quad (A) \quad and \quad \lim_{n \to \infty} n a_n = 0.$$

Then the series $\sum_{n=0}^{\infty} a_n$ *converges with a sum* S.

In fact, this theorem is the inverse to Abel's theorem and connects the Abel summability of a given numerical series and its ordinary summability, as well. It already has more than 100 years of history. Recently, Korevaar has published his survey paper [Korevaar (2002)] devoted to the Century Anniversary of complex Tauberian theory.

At first sight, it seems that the condition $a_n = o(1/n)$ is essential. Nevertheless, Littlewood succeeded to weaken it and obtained the following strengthened version of the Tauber theorem [Hardy (1949, Theorem 90)].

Theorem 3.2 (of Littlewood). *Let the series* (3.7) *be A-summable,*

$$\sum_{n=0}^{\infty} a_n = S \quad (A) \quad and \quad a_n = O(1/n).$$

Then the series $\sum_{n=0}^{\infty} a_n$ *converges with a sum* S.

Further, in order to discuss the boundary behavior of the power series (3.6), we need to first cite another definition. To this purpose, let $\{a_n\}_{n=0}^{\infty}$ be a sequence of complex numbers with

$$\limsup_{n \to \infty} (|a_n|)^{1/n} = R^{-1}, \quad 0 < R < \infty,$$

and $f(z)$ be the sum of the power series (3.6) in the open disk $D(0; R)$, i.e.

$$f(z) = \sum_{n=0}^{\infty} a_n z^n, \quad z \in D(0; R). \tag{3.8}$$

Definition 3.2. A point $z_0 \in \partial D(0; R)$ is called *regular* point for the function f if there exist a neighborhood $U(z_0; \rho)$ and a function $f_{z_0}^* \in \mathcal{H}(U(z_0; \rho))$ (the space of complex-valued functions, holomorphic in the set $U(z_0; \rho)$), such that $f_{z_0}^*(z) = f(z)$ for $z \in U(z_0; \rho) \cap D(0; R)$.

By this definition, it follows that the set of regular points of the power series is an open subset of the circle $C(0; R) = \partial D(0; R)$ with respect to the relative topology on $\partial D(0; R)$, i.e. the topology induced by that of $\mathbb{C}$.

In general, there is no relation between the convergence (divergence) of a power series at points on the boundary of its disk of convergence and the regularity (singularity) of its sum of such points. For example, the power series $\sum_{n=0}^{\infty} z^n$ is divergent at each point of the unit circle $C(0; 1)$ regardless of the fact that all the points of this circle, except for $z = 1$, are regular for its sum. The series $\sum_{n=1}^{\infty} n^{-2} z^n$ is (absolutely) convergent at each point of the circle $C(0; 1)$, but nevertheless one of them, namely $z = 1$, is a singular (i.e. not regular) for its sum. But under additional conditions on the sequence $\{a_n\}_{n=0}^{\infty}$, such a relation does exist (see for details [Markushevich (1967, Vol. 1, Chapter 3, p. 357, §7, 7.3)]), namely, as follows.

Theorem 3.3 (of Fatou). *If the coefficients of the power series with the unit disk of convergence tend to zero, i.e.* $\lim_{n\to\infty} a_n = 0$, *then the power series converges, even uniformly, on each arc of the unit circle, all points of which (including the ends of the arc) are regular for the sum of the series.*

Recall that it is possible for a given power series with a finite radius of convergence $0 < R < \infty$ to be convergent or divergent at some points of the boundary $C(0; R)$. These points could be regular or singular for its sum f, but the series diverges outside the domain of convergence. However, sometimes it is possible for a subsequence of its partial sums to exist that converges in the neighborhood of a regular point of the sum. In order to introduce the next definition ([Markushevich (1967, Vol. 2, p. 500)]), we first set

$$s_p(z) = \sum_{k=0}^{p} a_k z^k, \quad p = 0, 1, 2, \ldots. \tag{3.9}$$

Definition 3.3. A power series with a finite radius of convergence R is said to be *overconvergent*, if there exist a subsequence $\{s_{p_k}\}_{k=0}^{\infty}$ of the partial-sums sequence $\{s_p\}_{p=0}^{\infty}$ and a region G, containing the open disk $D(0; R)$, $G \cap \partial D(0; R) \neq \varnothing$, such that $\{s_{p_k}\}$ is uniformly convergent inside G.

Definition 3.4. We say that the function f (or the series), given by (3.8), possesses *Hadamard gaps*, if there exist two sequences $\{p_n\}_{n=0}^{\infty}$ and $\{q_n\}_{n=0}^{\infty}$, having the property $q_{n-1} \leq p_n \leq q_n/(1+\theta)$ $(\theta > 0)$ and $a_k = 0$ for $p_n < k < q_n$ $(n = 0, 1, 2, \dots)$.

Thus, beginning with the domain of convergence and series behavior near its boundary, passing through the possible uniform convergence on an arbitrary closed arc of the boundary, all the points of which are regular for its sum f, we come to the natural question: '*What type of conditions should be imposed on the power series that ensure the existence of subsequence* $\{s_{p_k}\}$, *convergent outside the disk of convergence?*'. The answer to this question is given in the early 20th century by Ostrowski [Ostrowski (1921, 1923)], see also [Kovacheva (2008)]. Namely, his classical result states that a given power series with existing regular points on the boundary of convergence disk is overconvergent iff it possesses Hadamard gaps. We draw the attention to the fact that merely the existence of Hadamard gaps does not imply overconvergence. For example, the power series $\sum_{n=0}^{\infty} a_{k_n} z^{k_n}$ with $k_{n+1} \geq (1+\theta)k_n$ $(\theta > 0)$ and $\limsup_{n\to\infty} (|a_{k_n}|)^{1/k_n} = 1$ possesses Hadamard gaps but nevertheless it is not overconvergent. Its natural boundary of analyticity is the unit circle $|z| = 1$ and that is nothing but the theorem about the gaps, belonging to Hadamard [1892].

3.3.   Series in Bessel Functions

Let $J_\nu(z)$ denote the Bessel functions, defined with (1.2). Let us consider the series, defined by means of the Bessel functions, of the type

$$\sum_{n=0}^{\infty} a_n J_n(z), \quad z \in \mathbb{C} \tag{3.10}$$

and briefly call them *Bessel series*.

In this chapter, we study their geometry of convergence, more precisely, we determine where these series converge and where they do not, and moreover, where the convergence is uniform and where it is not. Their disks of convergence have been found and their behavior on the boundaries of these domains have been studied, proving the theorems of Cauchy–Hadamard, Abel, Tauberian and Fatou types. The overconvergence of the Bessel series and Hadamard type theorem about the gaps are also established. The definitions and main statements, concerning the above-mentioned results, have first appeared in [Paneva-Konovska (1999, 2009, 2013a, 2015b)]. The asymptotic formulae, obtained for the Bessel functions in the cases of 'large' values of indices, are used for proving the convergence theorems for the considered series.

3.4. Cauchy–Hadamard Type Theorem

In this section, the domain of convergence of the Bessel series (3.10) is found and the corresponding Cauchy–Hadamard type theorem has been proved.

Theorem 3.4 (of Cauchy–Hadamard type). *The domain of convergence of the series* (3.10) *is the disk* $D(0; R)$ *with a radius of convergence*

$$R = 2 \left(\limsup_{n \to \infty} (|a_n|/\Gamma(n+1))^{1/n} \right)^{-1}. \qquad (3.11)$$

More precisely, the series (3.10) *is absolutely convergent in the disk* $D(0; R)$ *and divergent in the domain* $|z| > R$. *The cases* $R = 0$ *and* $R = \infty$ *are incorporated in the common case.*

Proof. For convenience, let us denote

$$u_n(z) = a_n J_n(z), \quad b_n = 2^{-1}(|a_n|/\Gamma(n+1))^{1/n},$$

$$\Lambda = 1/R = \limsup_{n \to \infty} b_n.$$

Using the asymptotic formula (1.19), we get

$$u_n(z) = a_n(z/2)^n(1 + \theta_n(z))/\Gamma(n + 1).$$

The proof goes in three cases.

(i) If $\Lambda = 0$, then $\lim_{n\to\infty} b_n = \limsup_{n\to\infty} b_n = 0$. Let us fix $z \neq 0$. Obviously, there exists a number N_1 such that for every $n > N_1$, the inequalities $|1 + \theta_n(z)| < 2$ and $2b_n < 1/|z|$ hold, whence $|u_n(z)| = b_n^n|z|^n|1 + \theta_n(z)| < 2^{1-n}$. The absolute convergence of (3.10) follows immediately from this inequality.

(ii) When $0 < \Lambda < \infty$. First, let z be in the domain $D(0; R)$, i.e. $|z|/R < 1$. Then $\limsup_{n\to\infty} |z|b_n < 1$. Therefore, a number $q < 1$ exists such that $\limsup_{n\to\infty} |z|b_n \leq q$, whence $|z|^n b_n^n \leq q^n$. Using the asymptotic formula for the general term $u_n(z)$ of the series (3.10), we obtain $|u_n(z)| = b_n^n|z|^n|1 + \theta_n(z)| \leq q^n|1 + \theta_n(z)|$. Since $\lim_{n\to\infty} \theta_n(z) = 0$, there exists N_2 such that $|1 + \theta_n(z)| < 2$ for every $n > N_2$, and hence $|u_n(z)| \leq 2q^n$. As the series $\sum_{n=0}^{\infty} 2q^n$ is convergent, the series (3.10) is also convergent, even absolutely.

Now, let z lie outside this domain, i.e. $|z|/R > 1$. Then $\limsup_{n\to\infty} |z|b_n > 1$ and therefore, there exist infinite number of values n_k of n with the property $|z|^{n_k} b_{n_k}^{n_k} > 1$. Since $\lim_{n\to\infty} \theta_n(z) = 0$, there exists N_3 so that for $n_k > N_3$, $|1 + \theta_{n_k}(z)| \geq 1/2$, i.e. $|u_{n_k}(z)| \geq 1/2$, for infinite number of values of n. This means that the necessary condition for convergence is not satisfied and therefore the series (3.10) is divergent.

(iii) When $\Lambda = \infty$. Let $z \in \mathbb{C}\backslash\{0\}$. Then $b_{n_k} > 1/|z|$ for infinite number of values n_k of n, whence $|u_{n_k}(z)| = |z|^{n_k} b_{n_k}^{n_k} |1 + \theta_{n_k}(z)| \geq 1/2$. In other terms, the necessary condition for the convergence of the series (3.10) is not satisfied, and we deduce that the series (3.10) is divergent for every $z \neq 0$. $\square$

Corollary 3.1. *Let the series* (3.10) *converge at the point* $z_0 \neq 0$. *Then it is absolutely convergent in the disk* $D(0; |z_0|)$. *Inside the disk*

$D(0; R)$, *i.e. on each closed disk* $|z| \le r$ $(r < R)$, *the convergence is uniform.*

Proof. So, since the considered series converges at the point $z_0 \ne 0$, then its radius of convergence R is a positive number, and moreover, the point z_0 lies either in the disk $D(0; R)$ or on its boundary — the circle $C(0; R)$. That is why, the disk $D(0; |z_0|)$ is either a part of the domain of convergence or it coincides with it, whence the absolute convergence follows. To prove uniformity of the convergence inside the disk $D(0; R)$, it is sufficient to show that the series is uniformly convergent on each closed disk $|z| \le r$ $(r < R)$. To this purpose, choosing a point ζ, $|\zeta| = \rho$, $r < \rho < R$, and considering the series (3.10), we estimate $|a_n J_n(z)|$. First, we mention that some of the values of $J_n(\zeta)$, but only finite numbers of them, can be zero. Then, bearing in mind (1.19), as well, there exists a number p, such that the expression $|a_n J_n(z)|$ can be written as follows:

$$
\begin{aligned}
|a_n J_n(z)| &= |a_n J_n(\zeta)| \frac{|J_n(z)|}{|J_n(\zeta)|} \\
&= |a_n J_n(\zeta)| \frac{|z^n||1 + \theta_n(z)|}{|\zeta^n||1 + \theta_n(\zeta)|} \\
&\le |a_n J_n(\zeta)| \frac{|1 + \theta_n(z)|}{|1 + \theta_n(\zeta)|},
\end{aligned}
$$

for all $n > p$ and $|z| \le r$.

Because of (1.23) and the relation $\lim_{n \to \infty} \frac{1}{n+1} = 0$, we obtain the equalities $\lim_{n \to \infty}(1 + \theta_n(z)) = 1$ and $\lim_{n \to \infty}(1 + \theta_n(\zeta))^{-1} = 1$. Therefore, there exists a number A such that $|1 + \theta_n(z)||1 + \theta_n(\zeta)|^{-1} \le A$ and hence $|a_n J_n(z)| \le A|a_n J_n(\zeta)|$, for all the values of $n > p$ and $|z| \le r$. Since the series $\sum_{n=0}^{\infty} a_n J_n(\zeta)$ is absolutely convergent and according to the Weierstrass criterium for uniform convergence, the proof is completed. □

The very disk of convergence is not obligatorily a domain of uniform convergence and on its boundary the series may even be divergent.

3.5. Abel Type Theorem

It turns out that Abel's theorem fails even for series of the kind $\sum_{k=1}^{\infty} a_{n_k} z^{n_k}$, where $(n_1, n_2, \ldots, n_k, \ldots)$ is a suitable permutation of the nonnegative integers [Tchakalov (1972, p. 92)]. Therefore, it is interesting to know if for the series in a given sequence of holomorphic functions, a statement like the Abel's theorem is available. A positive answer to this question, concerning the series in Laguerre and Hermite polynomials, is given by Rusev in his monographs [Rusev (1984, Chapter 11, §11.3); Rusev (2005, Chapter 4, §4)], and Boyadjiev [1986].

Let $z_0 \in \mathbb{C}$, $0 < R < \infty$, $|z_0| = R$ and g_φ be an arbitrary angular domain like (3.4) with a size $2\varphi < \pi$ and a vertex at the point $z = z_0$, that is symmetric with respect to the line passing through the points 0 and z_0, and d_φ be the part of the angular domain g_φ, closed between the angle's arms and the arc of the circle centered at the origin and touching the arms of the angle. The following theorem refers to the uniform convergence of the series (3.10) in the set d_φ and the limit of its sum at the point z_0, provided $z \in g_\varphi$.

Theorem 3.5 (of Abel type). *Let $\{a_n\}_{n=0}^{\infty}$ be a sequence of complex numbers, R be the positive number defined by* (3.11), *$F(z)$ be the sum of the series* (3.10) *in the domain $D(0; R)$, i.e.*

$$F(z) = \sum_{n=0}^{\infty} a_n J_n(z), \quad z \in D(0; R), \tag{3.12}$$

and this series converges at the point z_0 of the boundary of $D(0; R)$. Then,

(i) *the series* (3.10) *is uniformly convergent in the domain d_φ; and*
(ii) *the following relation holds:*

$$\lim_{z \to z_0} F(z) = \sum_{n=0}^{\infty} a_n J_n(z_0), \tag{3.13}$$

provided $z \in g_\varphi$.

Proof. (i) To prove the uniform convergence, we use the inequality (3.5) that is the crucial point of the proof.

So, let $z \in d_\varphi$. Setting

$$S_k(z) = \sum_{n=0}^{k} a_n J_n(z),$$

$$S_k(z_0) = \sum_{n=0}^{k} a_n J_n(z_0), \quad \lim_{k \to \infty} S_k(z_0) = s, \qquad (3.14)$$

$$\beta_n = S_n(z_0) - s, \quad \beta_n - \beta_{n-1} = a_n J_n(z_0),$$

we obtain

$$S_{k+p}(z) - S_k(z) = \sum_{n=0}^{k+p} a_n J_n(z) - \sum_{n=0}^{k} a_n J_n(z) = \sum_{n=k+1}^{k+p} a_n J_n(z).$$

According to Remark 1.2, there exists a natural number N_0 such that $J_n(z_0) \neq 0$ when $n > N_0$. Let $k > N_0$ and $p > 0$. Then, using the denotation

$$\gamma_n(z; z_0) = J_n(z)/J_n(z_0),$$

we can write the difference $S_{k+p}(z) - S_k(z)$ as follows:

$$S_{k+p}(z) - S_k(z) = \sum_{n=k+1}^{k+p} a_n J_n(z_0) \frac{J_n(z)}{J_n(z_0)} = \sum_{n=k+1}^{k+p} a_n J_n(z_0) \gamma_n(z; z_0).$$

Now, by Abel's transformation (see in [Markushevich (1967, Vol. 1, Chapter 1, p. 32, 3.4:7)]), we obtain subsequently:

$$S_{k+p}(z) - S_k(z) = \sum_{n=k+1}^{k+p} (\beta_n - \beta_{n-1}) \gamma_n(z; z_0)$$

$$= \beta_{k+p} \gamma_{k+p}(z; z_0) - \beta_k \gamma_{k+1}(z; z_0)$$

$$- \sum_{n=k+1}^{k+p-1} \beta_n (\gamma_{n+1}(z; z_0) - \gamma_n(z; z_0)),$$

$$S_{k+p}(z) - S_k(z) = (S_{k+p}(z_0) - s)\gamma_{k+p}(z; z_0)$$
$$- (S_k(z_0) - s)\gamma_{k+1}(z; z_0)$$
$$+ \sum_{n=k+1}^{k+p-1} (S_n(z_0) - s) \times \left(\frac{J_n(z)}{J_n(z_0)} - \frac{J_{n+1}(z)}{J_{n+1}(z_0)} \right).$$

So, using the last relation, we are going to estimate the module of the difference $S_{k+p}(z) - S_k(z)$ as follows:

$$|S_{k+p}(z) - S_k(z)|$$
$$\leq |S_{k+p}(z_0) - s||\gamma_{k+p}(z; z_0)| + |S_k(z_0) - s||\gamma_{k+1}(z; z_0)|$$
$$+ \sum_{n=k+1}^{k+p-1} |S_n(z_0) - s| \times \left| \frac{J_n(z)}{J_n(z_0)} - \frac{J_{n+1}(z)}{J_{n+1}(z_0)} \right|. \qquad (3.15)$$

Because of (1.23) and the relations $\lim_{n\to\infty} \frac{1}{n+1} = 0$ and $\lim_{n\to\infty}(1 + \theta_n(z_0))^{-1} = 1$, there exist numbers A and $N_1 > N_0$ such that $|1 + \theta_n(z)| \leq A/2$ for all the natural values of n and $|1 + \theta_n(\zeta)|^{-1} < 2$ for $n > N_1$, whence

$$|\gamma_n(z, z_0)| \leq A \quad \text{for } n > N_1. \qquad (3.16)$$

Further, setting

$$j_n(z, z_0) = \frac{J_n(z)}{J_n(z_0)} - \frac{J_{n+1}(z)}{J_{n+1}(z_0)}$$
$$= \frac{z^n}{z_0^n} \times \left(\frac{1 + \theta_n(z)}{1 + \theta_n(z_0)} - \frac{z}{z_0} \times \frac{1 + \theta_{n+1}(z)}{1 + \theta_{n+1}(z_0)} \right)$$

and observing that $j_n(z_0, z_0) = 0$, we apply the Schwarz lemma, named after Hermann Amandus Schwarz, for $j_n(z, z_0)$. Thus, we get that there exists a constant C:

$$|j_n(z, z_0)| = |J_n(z)/J_n(z_0) - J_{n+1}(z)/J_{n+1}(z_0)| \leq C|z - z_0||z/z_0|^n,$$

whence, and in accordance with (3.5),

$$\sum_{n=k+1}^{k+p+1} |j_n(z, z_0)| \leq \sum_{n=0}^{\infty} C|z - z_0||z/z_0|^n$$

$$= C|z_0| \times \frac{|z - z_0|}{|z_0| - |z|} < \frac{2C|z_0|}{\cos \varphi}. \tag{3.17}$$

Let ε be an arbitrary positive number. Taking in view the third of the relations (3.14), we can confirm that there exists a positive number $N_2 > N_0$ so large that

$$|S_n(z_0) - s| < \min\left(\frac{\varepsilon}{3A}, \frac{\varepsilon \cos \varphi}{6C|z_0|}\right) \quad \text{for } n > N_2. \tag{3.18}$$

Now, letting $N = N(\varepsilon) = \max(N_1, N_2)$ and $k > N$, the inequalities (3.15)–(3.18) give

$$|S_{k+p}(z) - S_k(z)| < \frac{2\varepsilon}{3} + \frac{\varepsilon \cos \varphi}{6C|z_0|} \times \sum_{n=k+1}^{k+p+1} |j_n(z, z_0)|$$

$$< \frac{2\varepsilon}{3} + \frac{\varepsilon \cos \varphi}{6C|z_0|} \times \frac{2C|z_0|}{\cos \varphi} = \varepsilon,$$

that completes the proof of (i).

(ii) The second part of the theorem could be proved in a similar way, estimating the module of the second summand of the difference

$$\Delta(z) = \sum_{n=0}^{\infty} a_n J_n(z_0) - F(z)$$

$$= \sum_{n=0}^{k} a_n(J_n(z_0) - J_n(z)) + \sum_{n=k+1}^{\infty} a_n(J_n(z_0) - J_n(z)),$$

'near' the vertex of the angular domain g_φ in the part d_φ. However, here we give another proof as a corollary of the first part. Namely, the uniform convergence of the series along with the equalities

$\lim_{z \to z_0} J_n(z) = J_n(z_0)$ $(n \in \mathbb{N}_0)$ verify the equality (3.13) that completes the proof for the considered series. $\square$

Remark 3.3. If the series (3.10) has a finite and nonzero radius of convergence R, also if it converges at the point $z_0 \in C(0; R)$ and F is the holomorphic function defined by this series in its domain of convergence $D(0; R)$, then according to the Theorem 3.5 it follows that

$$\lim_{z \to z_0, \, z \in d_\varphi} F(z) = F(z_0),$$

i.e. the restriction of the function F to each set of the kind d_φ is continuous at the point z_0.

3.6. (J, z_0)-Summation

Let $z_0 \in \mathbb{C} \setminus \mathbb{R}$ and $|z_0| = R$. Since all the zeros of $J_n(z)$ are real, then $J_n(z_0) \neq 0$. For the sake of brevity, denote

$$J_n^*(z; z_0) = \frac{J_n(z)}{J_n(z_0)}, \quad z \in \mathbb{C}. \tag{3.19}$$

Remark 3.4. We observe that all the functions of the family

$$J = \left\{ J_n^*(z; z_0), \, n = 0, 1, \ldots \right\}$$

are entire functions satisfying the condition $J_n^*(z_0; z_0) = 1$.

Definition 3.5. The numerical series (3.7) is said to be (J, z_0)-*summable* if the series

$$\sum_{n=0}^{\infty} a_n J_n^*(z; z_0) \tag{3.20}$$

converges in the disk $D(0; R)$ and, moreover, there exists the limit

$$\lim_{z \to z_0} \sum_{n=0}^{\infty} a_n J_n^*(z; z_0), \tag{3.21}$$

provided z remains on the segment $[0, z_0)$.

Remark 3.5. Every (J, z_0)-summation is regular, and this property is just a particular case of Theorem 3.5.

For convenience, in order to make this definition more universal and usable for various considerations, we intend to paraphrase the above definition. For this purpose, taking in view Remark 3.4, we firstly introduce one more denotation. Let $z_0 \in \mathbb{C}$, $z_0 \neq 0$, $|z_0| = R$, $0 < R < \infty$, and

$$(J; z_0) := \{j_n : j_n\text{-entire function}, j_n(z_0) = 1\}_{n \in \mathbb{N}_0}. \qquad (3.22)$$

Now, considering the series given below:

$$\sum_{n=0}^{\infty} a_n j_n(z), \quad j_n \in (J; z_0), \qquad (3.23)$$

Definition 3.5 can be rewritten as follows.

Definition 3.6. The numerical series (3.7) is said to be (J, z_0)-*summable*, if the series (3.23) converges in the disk $D(0; R)$, and moreover, there exists the limit

$$\lim_{z \to z_0} \sum_{n=0}^{\infty} a_n j_n(z), \qquad (3.24)$$

provided z remains on the segment $[0, z_0)$ (i.e. z radially tends to z_0).

However, using this definition must necessarily take into account of the regularity of the summation.

3.7. Tauber Type Theorem

As it has been emphasized in Sec. 3.2, the Tauber theorem is a statement that relates the Abel summability and the usual convergence of a numerical series by means of some assumptions imposed on the general term of the series under consideration. The classical result in this direction is cited there. Analogical result for summation by means of Laguerre polynomials is obtained by Rusev [1977a]. Here, we give a Tauber type theorem in a sense of the (J, z_0)-summation, defined by Definition 3.5.

To this purpose, let us consider the numerical series (3.7), $z_0 \in \mathbb{C} \setminus \mathbb{R}$, $|z_0| = R$, $0 < R < \infty$, and $J_n^*(z; z_0)$ be given by the relation (3.19). Let the series (3.10) converge for $z \in D(0; R)$ and

$$F(z) = \sum_{n=0}^{\infty} a_n J_n^*(z; z_0), \quad z \in D(0; R). \qquad (3.25)$$

The next theorem connects the (J, z_0)-summation of the numerical series (3.7) and its usual convergence, under additional condition on the growth of the general term a_n.

Theorem 3.6 (of Tauber type). *If the series (3.7) is (J, z_0)-summable and*

$$\lim_{n \to \infty} n a_n = 0, \qquad (3.26)$$

then it is convergent.

Proof. Let z belong to the segment $[0, z_0]$. By using the asymptotic formula (1.19) for the Bessel functions of the first kind, we obtain:

$$a_n J_n^*(z; z_0) = a_n \left(\frac{z}{z_0}\right)^n \frac{1 + \theta_n(z)}{1 + \theta_n(z_0)} = a_n \left(\frac{z}{z_0}\right)^n \left(1 + \widetilde{\theta}_n(z; z_0)\right),$$

where

$$\widetilde{\theta}_n(z; z_0) = \frac{\theta_n(z) - \theta_n(z_0)}{1 + \theta_n(z_0)}.$$

Then $\widetilde{\theta}_n(z; z_0) = O(1/n)$, due to (1.23).

Let us write (3.20) in the form:

$$\sum_{n=0}^{\infty} a_n J_n^*(z; z_0) = \sum_{n=0}^{\infty} a_n \left(\frac{z}{z_0}\right)^n (1 + \widetilde{\theta}_n(z; z_0)). \qquad (3.27)$$

Denoting

$$w_n(z) = a_n \left(\frac{z}{z_0}\right)^n \widetilde{\theta}_n(z; z_0), \qquad (3.28)$$

we consider the series $\sum_{n=0}^{\infty} w_n(z)$. According to the condition (3.26), the numerical sequence $\{n a_n\}_{n=0}^{\infty}$, being a convergent sequence, is bounded. Then, since $|w_n(z)| \le |a_n| |\widetilde{\theta}_n(z; z_0)|$ and bearing in mind

(1.23), there exists a constant C, such that $|w_n(z)| \leq C/n^2$ for all the positive integers n. Since $\sum_{n=1}^{\infty} 1/n^2$ converges, the series $\sum_{n=0}^{\infty} w_n(z)$ is also convergent, even absolutely and uniformly on the segment $[0, z_0]$. Therefore, by changing the order of the limit and summation, in view of the equality $\lim_{z \to z_0} w_n(z) = 0$, we deduce that

$$\lim_{z \to z_0} \sum_{n=0}^{\infty} w_n(z) = \sum_{n=0}^{\infty} \lim_{z \to z_0} w_n(z) = 0. \tag{3.29}$$

The assumption that the series (3.7) is (J, z_0)-summable implies the existence of the limit (3.21). Then, bearing in mind that (3.27) can be written in the form:

$$\sum_{n=0}^{\infty} a_n J_n^*(z; z_0) = \sum_{n=0}^{\infty} a_n \left(\frac{z}{z_0} \right)^n + \sum_{n=0}^{\infty} a_n \left(\frac{z}{z_0} \right)^n \widetilde{\theta}_n(z; z_0),$$

we conclude that the limit

$$\lim_{z \to z_0} \sum_{n=0}^{\infty} a_n \left(\frac{z}{z_0} \right)^n \tag{3.30}$$

also exists and, moreover, in view of (3.29),

$$\lim_{z \to z_0} \sum_{n=0}^{\infty} a_n J_n^*(z; z_0) = \lim_{z \to z_0} \sum_{n=0}^{\infty} a_n \left(\frac{z}{z_0} \right)^n. \tag{3.31}$$

From the existence of the limit (3.30), it follows that the series (3.7) is A-summable. Then according to Theorem 3.1, the series (3.7) converges. $\qquad \square$

A Littlewood generalization of the $o(1/n)$ version of the Tauber type theorem (Theorem 3.6) is given in this part, as well.

Theorem 3.7 (of Littlewood type). *If the series* (3.7) *is* (J, z_0)-*summable and*

$$a_n = O(1/n), \tag{3.32}$$

then the series (3.7) *converges.*

Proof. Let z belong to the segment $[0, z_0]$. In the beginning, consider the expression $w_n(z)$, given in (3.28). Using the conditions (3.32) and (1.23), we can again conclude that there exists a constant C, such that $|w_n(z)| \leq C/n^2$ for all the values of $n \in \mathbb{N}$. From this point, the proof follows the same line as that of Theorem 3.6, using Theorem 3.2 instead of Theorem 3.1. $\qquad\square$

3.8. Fatou Type Theorem

In order to obtain the simplest possible results, we need to select a suitable system of functions. To specify the family of Bessel functions in which we consider the series in the complex plane and their convergence, we modify the functions J_n a little bit, multiplying each of them with a corresponding coefficient. In what follows, we shall use the notation

$$\widetilde{J}_n(z) = n! 2^n J_n(z), \quad n = 0, 1, \ldots, \tag{3.33}$$

and consider the series

$$\sum_{n=0}^{\infty} a_n \widetilde{J}_n(z), \quad a_n \in \mathbb{C}, \ z \in \mathbb{C}, \tag{3.34}$$

that converges absolutely in the disk $D(0; R)$ with

$$R = \left(\limsup_{n \to \infty} (|a_n|)^{1/n} \right)^{-1},$$

and diverges in the domain $|z| > R$.

In studying the boundary behavior of the series (3.34) and considering for convenience the case $R = 1$, we give a relation between the distribution of the regular points of the function (3.35) on the circle $C(0; 1)$ and the uniform convergence of the series (3.34) on the set of such points. Propositions referring to the discussed property have also been established for the series in the Laguerre and Hermite polynomials (see e.g. [Rusev (2005)]). Here, we give such a type of theorem for the considered Bessel series (3.34), as follows.

Theorem 3.8 (of Fatou type). *Let* $\{a_n\}_{n=0}^{\infty}$ *be a sequence of complex numbers satisfying the conditions* $\lim_{n \to \infty} a_n = 0,$

$\limsup_{n\to\infty}(|a_n|)^{1/n} = 1$, *and* $F(z)$ *be the sum of the series* (3.34) *in the unit disk* $D(0;1)$, *i.e.*

$$F(z) = \sum_{n=0}^{\infty} a_n \widetilde{J}_n(z), \quad z \in D(0;1). \tag{3.35}$$

Let γ be an arbitrary arc of the unit circle $C(0;1)$ with all its points (including the ends) regular to the function F. Then the series (3.34) *converges, even uniformly, on the arc γ.*

Proof. Since all the points of the arc γ are regular to the function $F(z)$, there exists a region $G \supset \gamma$ where the function F can be continued. Denoting $\widetilde{G} = G \cup D(0;1)$, we define the function ψ in the region $\widetilde{G}$ satisfying the condition:

$$\psi(z) = F(z), \quad z \in D(0;1). \tag{3.36}$$

More precisely, it means that ψ is a single-valued analytical continuation of F into the domain $\widetilde{G}$.

Let $\rho > 0$ be the distance between the boundary $\partial\widetilde{G}$ of the region $\widetilde{G}$ and the arc γ ($\partial\widetilde{G}$ contains a part of the unit circle $|z| = 1$), and take the points ζ_1 and ζ_2:

$$\zeta_1, \zeta_2 \notin \gamma, \quad |\zeta_1| = |\zeta_2| = 1,$$

such that the distances between each of the points ζ_1 and ζ_2 and the respective proximal ends of the arc γ are equal to $\rho/2$ (see Fig. 3.3). Also, let the points $z_{1,2}$ and sector Δ be

$$z_1 = \zeta_1(1 + \rho/2), \quad z_2 = \zeta_2(1 + \rho/2), \quad \Delta = O z_1 z_2.$$

Define the auxiliary function

$$\varphi_n(z) = \psi(z) - \sum_{k=0}^{n} a_k \widetilde{J}_k(z) \tag{3.37}$$

and note that, according to Remark 1.2, there exists a natural number N_0 such that $\widetilde{J}_n(z) \neq 0$ when $z \neq 0$ and $n > N_0$. Now,

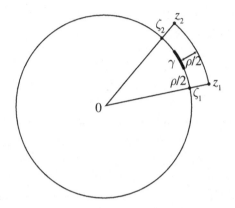

Fig. 3.3. The arc γ and sector $\Delta = Oz_1z_2$.

letting $n \geq N_0$, we introduce the notation

$$\omega_n(z) = \frac{\varphi_n(z)}{\widetilde{J}_{n+1}(z)}(z - \zeta_1)(z - \zeta_2), \quad z \neq 0; \quad \omega_n(0) = a_{n+1}\zeta_1\zeta_2.$$

$$(3.38)$$

In view of (3.35)–(3.38), and the convergence of the series (3.34) in the unit disk, as well, the function (3.38), constructed above, is holomorphic in a neighborhood of the origin. That is why it is holomorphic in the whole domain $\widetilde{G}$, where φ_n is holomorphic, too. For the estimation that we need, ω_n is preferred to φ_n. Namely, at the points of the boundary $\partial\Delta$ of the sector Δ, its denominator, containing $|z|^n$, becomes 'large' for 'large' values of n, but at the points of the radius, near $\zeta_{1,2}$, the corresponding differences $z - \zeta_{1,2}$ are 'small' modulo.

In order to prove that the sequence $\{\sum_{k=0}^{n} a_k \widetilde{J}_k(z)\}$ is uniformly convergent on the arc γ, it is sufficient to show that the sequence $\{\omega_n(z)\}_{n=N_0}^{\infty}$ uniformly tends to zero on the boundary $\partial\Delta$ of the sector $\Delta = Oz_1z_2$ and subsequently we estimate $\varphi_n(z)$ on the arc γ.

After these explanations, we are going to estimate the modulus $|\omega_n(z)|$ on the different parts of the boundary $\partial\Delta$. To this end, we come back to (1.23). We just mention that since $\lim_{n\to\infty} \frac{1}{n+1} = 0$, there exist numbers C and $\widetilde{N} > N_0$ such that $|1 + \theta_n(z)| \leq C/2$ for all the values of $n \in \mathbb{N}$ and $1/2 \le |1 + \theta_n(z)| \le 2$ for $n > \widetilde{N}$ on an arbitrary compact subset of $\mathbb{C}$.

Now, taking $\varepsilon > 0$ and setting

$$R = 1 + \rho/2, \quad \varepsilon_1 = \frac{\varepsilon \rho^3}{8(8CR^2 + \rho)},$$

$$M = \max_{z \in [\Delta]} |\psi(z)| \quad ([\Delta] = \Delta \cup \partial\Delta),$$

we have to consider four cases as follows.

(i) First, let $z \in (O, \zeta_1) \cup (O, \zeta_2) \subset D(0; 1)$.

In the unit disk, according to (3.37), we have consecutively:

$$\omega_n(z) = \sum_{k=0}^{\infty} a_{n+k+1} \frac{\widetilde{J}_{n+k+1}(z)}{\widetilde{J}_{n+1}(z)} (z - \zeta_1)(z - \zeta_2),$$

$$\omega_n(z) = \sum_{k=0}^{\infty} a_{n+k+1} \, z^k \frac{(1 + \theta_{n+k+1}(z))}{(1 + \theta_{n+1}(z))} (z - \zeta_1)(z - \zeta_2).$$

$$(3.39)$$

Since both $a_n \to 0$ and $|z - \zeta_1||z - \zeta_2| < 2(1 - |z|)$, there exists a number $N_1 = N_1(\varepsilon_1) > \widetilde{N}$, such that

$$|\omega_n(z)| < \varepsilon_1 \sum_{k=0}^{\infty} |z|^k \left| \frac{(1 + \theta_{n+k+1}(z))}{(1 + \theta_{n+1}(z))} \right| |(z - \zeta_1)||(z - \zeta_2)|$$

$$< 2C\varepsilon_1 \sum_{k=0}^{\infty} |z|^k (1 - |z|) = 2C\varepsilon_1$$

for $n > N_1$, i.e.

$$|\omega_n(z)| < 2C\varepsilon_1. \tag{3.40}$$

(ii) $z \in (\zeta_1, z_1) \cup (\zeta_2, z_2)$.

In this case, $|z - \zeta_1| = |z| - 1$, $|z - \zeta_2| \le |z| + |\zeta_2| < 2R$, and taking into account (1.19) and (3.37), we can write the following inequalities for the absolute value of $\omega_n(z)$:

$$\omega_n(z) = \frac{\psi(z) - \sum_{k=0}^{n} a_k \, z^k \, (1 + \theta_k(z))}{z^{n+1}(1 + \theta_{n+1}(z))}(z - \zeta_1)(z - \zeta_2),$$

namely

$$|\omega_n(z)| \leq \frac{M + \sum_{k=0}^{n} |a_k||z|^k|(1 + \theta_k(z))|}{|z|^{n+1}|(1 + \theta_{n+1}(z))|} \, 2R(|z| - 1)$$

$$< 2R\left(2M + \sum_{k=0}^{N_1} C|a_k|R^k\right)\frac{(|z| - 1)}{|z|^{n+1}}$$

$$+ 2\varepsilon_1 RC\frac{(|z| - 1)}{|z|^{n+1}} \sum_{k=N_1+1}^{n} |z|^k.$$

Furthermore, bearing in mind that, on the one hand:

$$\frac{(|z| - 1)}{|z|^{n+1}} < \frac{(|z| - 1)}{|z|^{n+1} - 1} = \frac{1}{|z|^n + \cdots + 1} < \frac{1}{n + 1},$$

and on the other hand:

$$\sum_{k=N_1+1}^{n} |z|^k = \frac{|z|^{n+1} - |z|^{N_1+1}}{(|z| - 1)} < \frac{|z|^{n+1}}{(|z| - 1)},$$

we conclude that

$$|\omega_n(z)| < \frac{2R}{n + 1}\left(2M + \sum_{k=0}^{N_1} C|a_k|R^k\right) + 2\varepsilon_1 RC.$$

Then, since $n^{-1} \to 0$, there exists a number $N_2 = N_2(\varepsilon_1) > N_1$, such that

$$\frac{2R}{n + 1}\left(2M + \sum_{k=0}^{N_1} C|a_k|R^k\right) < \varepsilon_1$$

for $n > N_2$, i.e.

$$|\omega_n(z)| < (1 + 2RC)\varepsilon_1. \tag{3.41}$$

(iii) z belongs to the arc $\widehat{z_1 z_2}$ (including the ends).
Then $|z - \zeta_1| < 2R$, $|z - \zeta_2| < 2R$ and hence, similarly to Case ii, the following estimates can be written, namely

$$|\omega_n(z)| < \frac{4R^2 \left(2M + \sum_{k=0}^{n} C |a_k| R^k \right)}{R^{n+1}}$$

$$< \frac{4 \left(2M + \sum_{k=0}^{N_1} C |a_k| R^k \right)}{R^{n-1}} + \frac{4\varepsilon_1 C \left(\sum_{k=N_1+1}^{n} R^k \right)}{R^{n-1}}$$

$$< \frac{4 \left(2M + \sum_{k=0}^{N_1} C |a_k| R^k \right)}{R^{n-1}} + \frac{8\varepsilon_1 C R^2}{\rho}.$$

Since $R^{-n} \to 0$, there exists a number $N_3 = N_3(\varepsilon_1) > N_1$, such that

$$|\omega_n(z)| < \left(\frac{8CR^2}{\rho} + 1 \right) \varepsilon_1 \qquad (3.42)$$

for $n > N_3$.

(iv) $z \in \{O, \zeta_1, \zeta_2\}$.
In this case, we have $\omega_n(0) = a_{n+1}\zeta_1\zeta_2$, whence $|\omega_n(0)| = |a_{n+1}| < \varepsilon_1$ for $n > N_1$, and $\omega_n(\zeta_{1,2}) = 0$, i.e.

$$|\omega_n(z)| < \varepsilon_1 \qquad (3.43)$$

for $n > N_1$.

Let $N = \max\{N_1, N_2, N_3\}$ and $n > N$, then bearing in mind the inequalities (3.40)–(3.43), we can write:

$$|\omega_n(z)| < \max \left(2C\varepsilon_1, (2RC + 1)\varepsilon_1, \left(\frac{8CR^2}{\rho} + 1 \right) \varepsilon_1 \right)$$

$$= \left(\frac{8CR^2}{\rho} + 1 \right) \varepsilon_1$$

on the boundary of the region Δ.

Therefore, in accordance with the principle of the maximum of the modulus,

$$|\omega_n(z)| < \left(\frac{8CR^2 + \rho}{\rho}\right)\varepsilon_1, \quad z \in \gamma. \tag{3.44}$$

Finally, according to (1.19), (3.37) and (3.39), since $|z| = 1$ on the arc γ, the following estimate can be produced:

$$
\begin{aligned}
|\omega_n(z)| &= \frac{\left|\psi(z) - \sum_{k=0}^n a_k \widetilde{J}_k(z)\right|}{|z^{n+1}||1 + \theta_{n+1}(z)|}|z - \zeta_1||z - \zeta_2| \\
&> \frac{1}{2} \cdot \frac{\rho^2}{4}\left|\psi(z) - \sum_{k=0}^n a_k \widetilde{J}_k(z)\right|,
\end{aligned}
\tag{3.45}
$$

and subsequently, applying both inequalities (3.44) and (3.45) yields

$$\left|\psi(z) - \sum_{k=0}^n a_k \widetilde{J}_k(z)\right| < \frac{8}{\rho^2}|\omega_n(z)| < \frac{8\varepsilon_1}{\rho^3}(8CR^2 + \rho) = \varepsilon, \quad z \in \gamma,$$

that proves the theorem. □

3.9. Overconvergence Theorems

To introduce the corresponding notions 'overconvergence' and 'gaps' for the series (3.34), through analogy with Definitions 3.3 and 3.4, the series (3.8) has to be replaced by the Bessel series (3.34) and, respectively, the sequence $\{s_p\}$, defined by (3.9), by the sequence $\{S_p\}$ with

$$S_p(z) = \sum_{k=0}^p a_k \widetilde{J}_k(z), \quad p = 0, 1, 2, \ldots. \tag{3.46}$$

In the other words, the following definitions can be given.

Definition 3.7. The Bessel series (3.34), with a finite radius of convergence R, is said to be *overconvergent*, if there exist a subsequence $\{S_{p_k}\}_{k=0}^\infty$ of the partial-sums sequence $\{S_p\}_{p=0}^\infty$ and a region G, containing the open disk $D(0; R)$, $G \cap \partial D(0; R) \neq \varnothing$, such that $\{S_{p_k}\}$ is uniformly convergent inside G.

Definition 3.8. We say that the function F (or the series), given by
(3.34), possesses *Hadamard gaps*, if there exist two sequences $\{p_n\}_{n=0}^\infty$
and $\{q_n\}_{n=0}^\infty$, having the property $q_{n-1} \le p_n \le q_n/(1+\theta)$ $(\theta > 0)$
and $a_k = 0$ for $p_n < k < q_n$ $(n = 0, 1, 2, \dots)$.

Remark 3.6. To introduce the corresponding notions 'overconvergence' and 'gaps' for the series in some arbitrary family of functions,
through analogy with Definitions 3.7 and 3.8, the sequence $\{S_p\}$,
defined by (3.46), and its subsequence $\{S_{p_k}\}$ have to be replaced by
the corresponding sequence of partial sums $\{S_p\}$ and its subsequence.

So, we are going to formulate the following theorems, referring
to the Bessel series with Hadamard gaps.

Theorem 3.9 (about the overconvergence). *Let $\{a_n\}_{n=0}^\infty$ be a
sequence of complex numbers satisfying the condition* $\limsup_{n\to\infty}$
$(|a_n|)^{1/n} = 1$, *let $F(z)$ be the sum of the series* (3.34) *in the unit
disk $D(0;1)$, $F(z)$ have at least one regular point, belonging to the
circle $C(0;1)$, and let $F(z)$ possess Hadamard gaps. Then the series*
(3.34) *is overconvergent.*

Proof. Without loss of generality, we suppose that the point $z_0 = 1$
is regular to the function F. This means that F is analytically
continuable into a neighborhood U of the point 1.

Denoting $\widetilde{U} = U \cup D(0;1)$, we define the function ψ in the region
$\widetilde{U}$ by the equality

$$\psi(z) = F(z), \quad z \in D(0;1).$$

More precisely, it means that ψ is a single-valued analytical continuation of F into the domain $\widetilde{U}$.

Letting $\theta > 0$ and taking $\{p_n\}_{n=0}^\infty$, $\{q_n\}_{n=0}^\infty$ with the properties
$q_n \ge (1+\theta)p_n$, $p_n \ge q_{n-1}$ and $a_k = 0$ for $p_n < k < q_n$ $(n = 0, 1, 2, \dots)$, we define the auxiliary function

$$\varphi_n(z) = \psi(z) - S_{p_n} = \psi(z) - \sum_{k=0}^{p_n} a_k \widetilde{J}_k(z). \tag{3.47}$$

In order to prove that the sequence $\{S_{p_n}\}$ is uniformly convergent inside the region $\widetilde{U}$, we are going to apply the Hadamard theorem for the three disks [Markushevich (1967, Vol. 2)]. To this end, taking δ and ω in such a way that $0 < \omega < \delta < 1/2$, we consider the three circles C_1, C_2 and C_3, centered at the point $1/2$ and having respectively the radii $1/2 - \delta$, $1/2 + \omega$ and $1/2 + \delta$, such that $C_3 \subset \widetilde{U}$ and subsequently set

$$M_{n,j} = \max_{z \in C_j} |\varphi_n(z)|, \quad j = 1, 2, 3; \quad M = \max_{z \in C_3} |\psi(z)|.$$

Before evaluating $|\varphi_n(z)|$ we come back to (1.23). We just mention that since $\lim_{n \to \infty} \frac{1}{n+1} = 0$, there exists a number B such that $|1 + \theta_n(z)| \leq B$ for all the values of $n \in \mathbb{N}$ on an arbitrary compact subset of $\mathbb{C}$. Now, letting $0 < \eta < \delta/2$ implies the existence of $A = A(\eta)$, such that $|a_k| \leq AB^{-1}(1 - \eta)^{-k}$. To find an upper estimation of $|\varphi_n(z)|$, we intend to consider three different cases.

(i) First, let $z \in C_1 \subset D(0; 1)$. In the unit disk, according to (3.47), we have

$$\varphi_n(z) = \sum_{k=q_n}^{\infty} a_k \widetilde{J}_k(z).$$

Therefore,

$$|\varphi_n(z)| \leq \sum_{k=q_n}^{\infty} |a_k \widetilde{J}_k(z)|$$

$$= \sum_{k=q_n}^{\infty} |a_k z^k (1 + \theta_k(z))|$$

$$= \sum_{k=q_n}^{\infty} |a_k| |1 + \theta_k(z)| |z^k|$$

$$\leq A \sum_{k=q_n}^{\infty} (1 - \eta)^{-k} (1 - \delta)^k$$

$$= A \left(1 - \frac{1-\delta}{1-\eta}\right)^{-1} \left(\frac{1-\delta}{1-\eta}\right)^{q_n},$$

whence

$$M_{n,1} = O\left(\left(\frac{1-\delta}{1-\eta}\right)^{q_n}\right) = O\left(\left(\frac{1-\delta}{1-\eta}\right)^{(1+\theta)p_n}\right). \quad (3.48)$$

(ii) Now, let $z \in C_3$. In this case,

$$|\varphi_n(z)| = |\psi(z) - S_{p_n}| = \left|\psi(z) - \sum_{k=0}^{p_n} a_k \widetilde{J}_k(z)\right|$$

$$\leq |\psi(z)| + \sum_{k=0}^{p_n} |a_k \widetilde{J}_k(z)| \leq M + \sum_{k=0}^{p_n} |a_k||1 + \theta_k(z)||z^k|$$

$$\leq M + A \sum_{k=0}^{p_n} \left(\frac{1+\delta}{1-\eta}\right)^k = O\left(\left(\frac{1+\delta}{1-\eta}\right)^{p_n}\right),$$

and therefore,

$$M_{n,3} = O\left(\left(\frac{1+\delta}{1-\eta}\right)^{p_n}\right). \quad (3.49)$$

(iii) Finally, let $z \in C_2$. Then, in view of (3.48) and (3.49) and according to the Hadamard theorem for the three disks (for details see [Markushevich (1967, Vol. 2, formula (3.2:2))]), we can write

$$M_{n,2} = O\left(\left(\left(\frac{1-\delta}{1-\eta}\right)^{(1+\theta)\ln\frac{1+2\delta}{1+2\omega}}\left(\frac{1+\delta}{1-\eta}\right)^{\ln\frac{1+2\omega}{1-2\delta}}\right)^{p_n}\right).$$

$$(3.50)$$

Note that the limit of the inner part of (3.50), namely

$$\left(\frac{1-\delta}{1-\eta}\right)^{(1+\theta)\ln\frac{1+2\delta}{1+2\omega}}\left(\frac{1+\delta}{1-\eta}\right)^{\ln\frac{1+2\omega}{1-2\delta}},$$

is equal to

$$a = (1 - \delta)^{(1+\theta)\ln(1+2\delta)}(1 + \delta)^{-\ln(1-2\delta)} \qquad (3.51)$$

when ω and η tend to zero. Moreover, if δ tends to zero then $a < 1$. Therefore, taking the logarithm of a, we have

$$
\begin{aligned}
\ln a &= (1 + \theta)\ln(1 + 2\delta)\ln(1 - \delta) - \ln(1 - 2\delta)\ln(1 + \delta) \\
&= (1 + \theta)(2\delta + O(\delta^2))(-\delta + O(\delta^2)) \\
&\quad - (-2\delta + O(\delta^2))(\delta + O(\delta^2)) \\
&= (1 + \theta)(-2\delta^2 + O(\delta^3)) + 2\delta^2 + O(\delta^3) \\
&= -2\theta\delta^2 + O(\delta^3).
\end{aligned}
$$

Therefore, $\ln a < 0$ when $\delta \to 0$ and for this reason $a < 1$ if δ tends to zero.

That is why, $\lim_{n \to \infty} M_{n,2} = 0$. Additionally, according to Corollary 3.1, $\{S_p\}$ uniformly converges inside the disk $D(0; 1)$. For these two reasons, the series $\{S_{p_n}\}$ is uniformly convergent inside the whole region $\widetilde{U}$.   □

Theorem 3.10 (of Hadamard type about the gaps). *Let $\{a_k\}_{k=0}^{\infty}$ be a sequence of complex numbers satisfying the conditions as follows:* $\limsup_{n \to \infty} (|a_{k_n}|)^{1/k_n} = 1$ *for* $k_{n+1} \geq (1 + \theta)k_n$ $(\theta > 0)$, *and* $a_k = 0$ *for* $k_n < k < k_{n+1}$. *Let $F(z)$ be the sum of the series* (3.34) *in the unit disk $D(0; 1)$, i.e.*

$$F(z) = \sum_{n=0}^{\infty} a_{k_n} \widetilde{J}_{k_n}(z), \quad z \in D(0; 1).$$

Then all the points of the unit circle $C(0; 1)$ are singular for the function F, i.e. the unit circle is a natural boundary of analyticity for the series

$$\sum_{n=0}^{\infty} a_{k_n} \widetilde{J}_{k_n}(z). \qquad (3.52)$$

Proof. Let $|z_0| = 1$ and z_0 be regular for F. Denoting $p_n = k_n$ and $q_n = k_{n+1}$, we can conclude that $q_{n-1} \leq p_n \leq q_n/(1 + \theta)$ $(\theta > 0)$ and $S_{p_n} = \sum_{s=0}^{n} a_{k_s} \widetilde{J}_{k_s}(z)$. Therefore, analogously to the proof of Theorem 3.9, S_{p_n} uniformly converges in a neighborhood of z_0. But the radius of convergence of the series (3.52) is $R = 1$ and we come to contradiction. $\qquad \square$

Chapter 4

Bessel and Neumann Expansions

4.1. Neumann Series in the Complex Plane

Let $O_n(z)$ and $A_{n,\nu}(z)$ be respectively the Neumann polynomials, defined by (1.14) and (1.15), and their Gegenbauer generalizations (1.16). Let

$$\varepsilon_0 = 1 \quad \text{and} \quad \varepsilon_n = 2 \quad \text{for} \quad n = 1, 2, 3, \dots . \qquad (4.1)$$

Introducing the following denotation:

$$\widetilde{A}_{n,\nu}(z) = \begin{cases} \varepsilon_n O_n(z) & \text{for } \nu = 0 \\ A_{n,\nu}(z) & \text{for } \nu \neq 0 \end{cases} \quad (\nu \neq -1, -2, -3, \dots), \qquad (4.2)$$

we consider a series of the kind

$$\sum_{n=0}^{\infty} a_n \widetilde{A}_{n,\nu}(z), \quad z \in \mathbb{C}, \ a_n \in \mathbb{C}, \qquad (4.3)$$

and briefly call them Neumann series. First, we determine the domain of convergence of such a kind of series, giving analog of the classical Cauchy–Hadamard theorem for the power series. Following this, we continue with other results, analogical to the classical ones, including Laurent's theorem, analytical continuation of holomorphic function and many others. The chapter is mainly written on the basis of [Paneva-Konovska (2003)].

4.2. Cauchy–Hadamard Type Theorem

Denote with $D^\star(0; R)$ the circular domain

$$D^\star(0; R) = \big\{ z : z \in \overline{\mathbb{C}}, \quad |z| > R \big\}, \tag{4.4}$$

where $\overline{\mathbb{C}}$ means the extended complex plane $\mathbb{C} \cup \{\infty\}$, i.e. $D^\star(0; R)$ is the open 'disk' with a center at infinity.

Both an analog to the classical Cauchy–Hadamard theorem and a corollary of the Abel's lemma type concerning the series (4.3), are given below.

Theorem 4.1 (of Cauchy–Hadamard type). *The region of convergence of the series* (4.3) *is the 'disk' $D^\star(0; R)$ with a radius of convergence*

$$R = 2 \limsup_{n \to \infty} (|a_n||\Gamma(n + \nu + 1)|)^{\frac{1}{n}}. \tag{4.5}$$

More precisely, the series (4.3) *is absolutely convergent in the 'disk' $D^\star(0; R)$ and divergent in the domain $0 < |z| < R$. The cases $R = 0$ and $R = \infty$ fall in the general case.*

Corollary 4.1. *Let the series* (4.3) *converge at the point $z_0 \neq 0$. Then it is absolutely convergent in the disk $D^\star(0; |z_0|)$. Inside the disk $D^\star(0; R)$, i.e. on each closed disk $|z| \geq \widetilde{R}$ (with $\widetilde{R} > R$), the convergence is uniform.*

Proof. Taking into account the asymptotic representations (1.21) and (1.22), respectively, for $O_n(z)$ and $A_{n,\nu}(z)$, the proofs of the statements, discussed above, go by analogy of those of Theorem 3.4 and Corollary 3.1. We omit the details. $\square$

4.3. Laurent Type Theorem

An important property of functions holomorphic in a circular ring is their ability to be expanded in Laurent's series. A natural question is whether it is possible to have a Laurent type theorem about the expansion in a given system of holomorphic functions. Numerous results in this direction have been obtained by Schäfke in a series of papers [Schäfke (1960, 1961, 1963)] and such a generalization is made in his paper [Schäfke (1961, p. 164, Theorem 5)]. He succeeds in proving that if $g(z)$ is holomorphic in the ringlike set

$$D_{r,R} = \{z : z \in \mathbb{C}, \ 0 \le r < |z| < R \le \infty\}, \tag{4.6}$$

it can be expanded in a series of the type:

$$g(z) = \sum_{n=0}^{\infty} a_n J_n(z) + \sum_{n=0}^{\infty} b_n O_n(z), \tag{4.7}$$

which is absolutely and uniformly convergent on the compact subsets of the ring. The coefficients are expressed by the formulae:

$$a_n = \frac{\varepsilon_n}{2\pi i} \oint g(z) O_n(z) dz, \quad b_n = \frac{\varepsilon_n}{2\pi i} \oint g(z) J_n(z) dz, \tag{4.8}$$

with ε_n $(n = 0, 1, 2, 3, \dots)$, like in (4.1). Moreover, if $g(z)$ is holomorphic in a neighborhood of zero then $b_n = 0$, $n = 0, 1, 2, \dots$, and the series $g(z) = \sum_{n=0}^{\infty} a_n J_n(z)$ is also absolutely and uniformly convergent in the neighborhood of $z = 0$. If $g(z)$ is holomorphic in a neighborhood of $z = \infty$ and $g(\infty) = 0$, then $a_n = 0$, $n = 0, 1, 2, \dots$, and the series $g(z) = \sum_{n=0}^{\infty} b_n O_n(z)$ is absolutely and uniformly convergent in the neighborhood of $z = \infty$, as well.

The possibility to expand a function, holomorphic in a ring, in a series in the systems (4.2), defined for $\nu \ne -1, -2, \dots$, and $\{z^{-\nu} J_{n+\nu}(z)\}_{n=0}^{\infty}$ is studied in this section.

Let $D(0; R)$ and $D^{\star}(0; r)$ be the circular domains, defined by (3.1) and (4.4), respectively. The following theorem holds.

Theorem 4.2 (of Laurent type). *Let $F(z)$ be a holomorphic function in the ring $D_{r,R}$, given by (4.6). Let C and c be the positive oriented curves $|z| = R_1$ and $|z| = r_1$ $(r < r_1 < R_1 < R)$, respectively. Then $F(z)$ can be expanded in a series of the kind:*

$$F(z) = \sum_{n=0}^{\infty} \tilde{a}_n \tilde{A}_{n,\nu}(z) + \sum_{n=0}^{\infty} b_n z^{-\nu} J_{n+\nu}(z), \quad \nu \ne -1, -2, \dots,$$

$$\tag{4.9}$$

with coefficients

$$\widetilde{a}_n = \frac{1}{2\pi i} \int_c F(\zeta)\zeta^{-\nu}J_{n+\nu}(\zeta)d\zeta, \quad b_n = \frac{1}{2\pi i} \int_C F(\zeta)\widetilde{A}_{n,\nu}(\zeta)d\zeta.$$

$$(4.10)$$

It is absolutely and uniformly convergent on the compact subsets of the ring. In particular, if $F(z)$ is holomorphic in the disk $D(0;R)$, the coefficients $\widetilde{a}_n$ are equal to zero; if $F(z)$ is holomorphic in the 'disk' $D^\star(0;r)$ and $F(\infty) = 0$, the coefficients b_n are equal to zero.

Proof. Bearing in mind the denotation (4.2) and identities for the Cauchy kernel $1/(\zeta - z)$ [Watson (1949, 9.2 (2), 9.14)]:

$$\frac{z^\nu}{\zeta - z} = \sum_{n=0}^{\infty} A_{n,\nu}(\zeta)J_{n+\nu}(z), \quad |z| < |\zeta|,$$

$$\frac{1}{\zeta - z} = \sum_{n=0}^{\infty} \varepsilon_n O_n(\zeta)J_n(z), \quad |z| < |\zeta|,$$

we obtain

$$\frac{1}{\zeta - z} = \sum_{n=0}^{\infty} \widetilde{A}_{n,\nu}(\zeta)z^{-\nu}J_{n+\nu}(z), \quad |z| < |\zeta|.$$

Then if the point z is situated in the ring $r_1 < |z| < R_1$, we can write:

$$2\pi i F(z) = \int_C \frac{F(\zeta)}{\zeta - z}d\zeta + \int_c \frac{F(\zeta)}{z - \zeta}d\zeta$$

$$= \int_C \left(\sum_{n=0}^{\infty} F(\zeta)\widetilde{A}_{n,\nu}(\zeta)z^{-\nu}J_{n+\nu}(z) \right) d\zeta$$

$$+ \int_c \left(\sum_{n=0}^{\infty} F(\zeta)\zeta^{-n}J_{n+\nu}(\zeta)\widetilde{A}_{n,\nu}(z) \right) d\zeta. \quad (4.11)$$

From the explicit form of $J_{n+\nu}(z)$, given by (1.2), it is easy to see that

$$z^{-\nu} J_{n+\nu}(z) = \frac{z^n}{2^{n+\nu}\Gamma(n+\nu+1)}(1+\theta_n(z)); \quad \lim_{n\to\infty}\theta_n(z) = 0,$$

for $z \in \mathbb{C}$.

In view of the denotation (4.2) and asymptotic formulae (1.21) and (1.22) for $O_n(z)$ and $A_{n,\nu}(z)$, respectively, we get the asymptotic formula:

$$\widetilde{A}_{n,\nu}(z) = \frac{2^{n+\nu}\Gamma(n+\nu+1)}{z^{n+1}}(1+\Phi_n(z)); \quad \lim_{n\to\infty}\Phi_n(z) = 0.$$

Now, denoting

$$u_n(z,\zeta) = \widetilde{A}_{n,\nu}(\zeta)z^{-\nu}J_{n+\nu}(z), \quad z,\zeta \in \mathbb{C},$$

and taking into account the last formula, $u_n(z,\zeta)$ can be written as follows:

$$u_n(z,\zeta) = \frac{z^n}{\zeta^{n+1}}(1+\theta_n(z))(1+\Phi_n(\zeta));$$

$$\lim_{n\to\infty}\theta_n(z) = 0, \ \lim_{n\to\infty}\Phi_n(\zeta) = 0.$$

Using this formula, we conclude that there exists a constant M such that

$$|F(\zeta)u_n(z,\zeta)| \leq M\left(\frac{|z|}{R_1}\right)^n, \quad |\zeta| = R_1,$$

$$|F(\zeta)u_n(\zeta,z)| \leq M\left(\frac{r_1}{|z|}\right)^n, \quad |\zeta| = r_1.$$

Then, since the obtained series

$$\sum_{n=0}^{\infty}\left(\frac{|z|}{R_1}\right)^n \quad \text{and} \quad \sum_{n=0}^{\infty}\left(\frac{r_1}{|z|}\right)^n$$

are convergent, the series in (4.11) are absolutely and uniformly convergent along the curves C and c, respectively. That is why the change of the positions of integration and summation in (4.11) is correct. Therefore,

$$2\pi i F(z) = \sum_{n=0}^{\infty} \left(\int_C F(\zeta) \widetilde{A}_{n,\nu}(\zeta) d\zeta \right) z^{-\nu} J_{n+\nu}(z)$$

$$+ \sum_{n=0}^{\infty} \left(\int_c F(\zeta) \zeta^{-n} J_{n+\nu}(\zeta) d\zeta \right) \widetilde{A}_{n,\nu}(z).$$

The last obtained identity proves that the function $F(z)$ is represented in the form (4.9) with coefficients (4.10). The uniformity and absoluteness of the convergence on the compact subsets of the ring $D_{r,R}$ follow from the Theorems 3.4 and 4.1 and Corollaries 3.1 and 4.1. The validity of the rest follows immediately from (4.10).

$\square$

Remark 4.1. In particular, if $\nu = 0$, the representation (4.9) is reduced to (4.7) with coefficients (4.8).

4.4. Classical Results Related to Singularities and Analytical Continuation of Holomorphic Functions

'What can be said about the singularities of the holomorphic functions defined with a convergent power series if the coefficients are known?' The above problem attracted the attention of the mathematicians back in the end of the 19th century. Hadamard, Borel and many other great mathematicians dealt with it. A large part of the corresponding results are connected with — George Pólya, the famous Hungarian mathematician. At a later stage, Ludwig Bieberbach handled the difficult task of reviewing the publications devoted to this problem, resulting in the monograph [Bieberbach (1955)].

Taking into consideration the analogy with the known theorems of the theory of power series, Hille [1939, 1940] paid special attention to the problem of analytical continuation and determining the singular points of functions, defined by convergent series in Hermite functions. The existence of certain parallelism between the properties of the power series and the series in classical orthogonal polynomials, as well as the possibility to make an analogy, was established by Rusev [1984, 2005]. Namely, he proved that the well-known Ostrowski's 'overconvergence' theorem [Ostrowski (1921)] and Hadamard 'lacunae' theorem [Hadamard (1892, 1893)] hold for the series in Laguerre and Hermite polynomials. Boyadjiev [1990] generalized the Cowling's theorem [Cowling (1958)] for the series in Laguerre polynomials with a parameter α under the most possible generalized assumptions for this parameter.

A number of references considered, see e.g. [Bieberbach (1955); Mandelbrojt (1927)], it is emphasize that the question of the analytic continuation of a function, holomorphic in a neighborhood of the origin, is connected with the properties of the coefficients of the power series, that defines this function. Our interest in similar methods of investigation of the analytic continuation of functions, that are represented by Bessel series, is due to the research work of Nehari [1956]. As a result, we are certain that a detailed study and complete mastering of the similarity with the theory of power series will have a stimulating effect on the further development of the Bessel series theory.

Closely related with the analytical continuation of a holomorphic function is the notion 'main star' of the function.

Definition 4.1. The domain S, containing the origin, is said to be a *star* or a *starlike domain*, centered at the origin [Bieberbach (1955), 1.4], if its intersection with each straight line, passing through the point $z = 0$, is only one line segment. The starlike domain S_f is said to be *Mittg-Leffler main star* at the point $z = 0$ of a function $f(z)$, regular at this point, if S_f is the largest star centered at the origin, where the function $f(z)$ is still regular.

In fact, the main star S_f can be considered as a set of all the line segments beginning at the origin, along which the function $f(z)$ can be continued analytically, i.e. it is the set of all the points z in the complex plane such that f can be continued analytically along the line segment joining 0 and z. Normally, it contains the open disk of convergence $D(0; R)$, i.e. $S_f \supset D(0; R)$, also it is an open star-convex set (with respect to the point 0).

In relation to the singularities, we are going to formulate two basic theorems depending on those where the coefficients of the series, representing the function, are situated. For this purpose, and also for convenience, we introduce some denotations as follows.

Let $a_0, a_1, a_2, \ldots, a_n, \ldots$ be a sequence of complex numbers, such that

$$\left(\limsup_{n \to \infty} (|a_n|)^{\frac{1}{n}} \right)^{-1} = R > 0, \tag{4.12}$$

and $f(z)$ be defined by the expansion:

$$f(z) = \sum_{n=0}^{\infty} a_n z^n, \quad |z| < R. \tag{4.13}$$

Let A_θ denote an arbitrary angle domain with a size $\theta < \pi$ and a vertex at the point $z = 0$.

Now, if the number $R < \infty$ and the coefficients of the given power series are nonnegative or, more generally, they lie in an angular domain with a size less than π, then the point $z_0 = R$ is a singular point of the function $f(z)$, i.e. the following Pringsheim's theorem (for power series) [Tchakalov (1972, §60, Theorem 3)] holds.

Theorem 4.3 (of Pringsheim). *If the terms of the complex sequence $\{a_n\}_{n=0}^{\infty}$, under the condition (4.12), are situated in the domain A_θ ($\theta < \pi$) and $R < \infty$, then $z_0 = R$ is a singular point of the function $f(z)$.*

However, if the coefficients lie in the whole right half-plane, then we need some more additional conditions for guaranteeing the singularity of point $z_0 = R$. This fact is established by the classical Vivanti–Dienes theorem [Bieberbach (1955, 1.8.2)], as follows.

Theorem 4.4 (of Vivanti–Dienes). *If the terms of the sequence* $\{a_n\}_{n=0}^{\infty}$ *satisfy the condition* (4.12), *and additionally*

$$a_n' = \Re(a_n) \geq 0 \quad (n = 0, 1, 2, \ldots) \quad and \quad \left(\limsup_{n \to \infty} \left(|a_n'|\right)^{\frac{1}{n}}\right)^{-1}$$
$$= R < \infty,$$

then the point $z_0 = R$ *is singular for the function* $f(z)$.

Now, let J_n and I_n ($n = 0, 1, 2, \ldots$) be respectively the Bessel and modified Bessel functions of the first kind and with an index n. Along with the function $f(z)$, we also consider $g(z)$ and $\tilde{g}(z)$ that are defined by the expansions:

$$g(z) = \sum_{n=0}^{\infty} a_n 2^n \Gamma\left(n + \frac{1}{2}\right) J_n(z), \quad |z| < R, \qquad (4.14)$$

$$\tilde{g}(z) = \sum_{n=0}^{\infty} a_n 2^n \Gamma\left(n + \frac{1}{2}\right) I_n(z), \quad |z| < R, \qquad (4.15)$$

and they have the same radius R of convergence, as it is not hard to be verified.

Furthermore, the analytical continuations of the functions $f(z)$, $g(z)$ and $\tilde{g}(z)$ outside the disk of convergence of their series expansions are studied and relations between their main stars are found. Pringsheim's type theorem and Vivanti–Dienes type theorem are proved for the functions (4.14) and (4.15). Similar problems have also been considered by Bieberbach [1955], Nehari [1956] and Boyadjiev [1990].

4.5. Relation between the Main Stars of $f(z)$, $g(z)$ and $\tilde{g}(z)$

Seeking an analogy with the theory of power series, we give some results about the existing relations between the main stars of $f(z)$, $g(z)$ and $\tilde{g}(z)$.

Lemma 4.1. *The following integral relation between the functions* $f(z)$ *and* $g(z)$ *holds:*

$$g(z) = \frac{2}{\sqrt{\pi}} \int_0^1 \frac{f\left(z(1-t^2)\right)}{\sqrt{1-t^2}} \cos(zt)dt, \qquad (4.16)$$

for $|z| < R$.

Proof. Using the classical Poisson integral representation (1.17), applied for integer nonnegative values of the parameter ν,

$$J_n(z) = \frac{z^n}{\sqrt{\pi}2^n\Gamma\left(n+\frac{1}{2}\right)} \int_{-1}^1 \left(1-t^2\right)^{n-\frac{1}{2}} \exp(izt)dt, \qquad (4.17)$$

and putting it in the expansion (4.14), we get

$$g(z) = \sum_{n=0}^{\infty} \frac{a_n}{\sqrt{\pi}} z^n \int_{-1}^1 \left(1-t^2\right)^{n-\frac{1}{2}} \exp(izt)dt.$$

Since $\left|\left(1-t^2\right)^{n-\frac{1}{2}}e^{izt}\right| \leq e^R$ for $t \in [-1,1], n = 1,2,\ldots$, the convergence of the expansion $\sum_{n=1}^{\infty} a_n z^n \left(1-t^2\right)^{n-\frac{1}{2}} \exp(izt)$ is uniform with respect to t in this interval. Then we can change the order of the integration and summation, i.e.

$$g(z) = \frac{1}{\sqrt{\pi}} \int_{-1}^1 \left(\sum_{n=0}^{\infty} a_n \left(z\left(1-t^2\right)\right)^n\right) \frac{\exp(izt)}{\sqrt{1-t^2}} dt.$$

From the inequality $\left|z\left(1-t^2\right)\right| < R$ and equality (4.13), we get

$$g(z) = \frac{1}{\sqrt{\pi}} \int_{-1}^1 \frac{f\left(z\left(1-t^2\right)\right)}{\sqrt{1-t^2}} \exp(izt)dt,$$

from which the validity of (4.16) is seen immediately.   □

Lemma 4.2. *The functions* $f(z)$ *and* $\widetilde{g}(z)$ *are related by the integral*

$$\widetilde{g}(z) = \frac{2}{\sqrt{\pi}} \int_0^1 \frac{f\left(z\left(1-t^2\right)\right)}{\sqrt{1-t^2}} \cosh(zt)dt, \qquad (4.18)$$

for $|z| < R$.

Proof. Bearing in mind that [Erdélyi *et al.* (1953, 7.2 (12))]

$$I_n(z) = (-i)^n J_n(iz) \tag{4.19}$$

and using the formula (4.17), we get for $\widetilde{g}(z)$ consequently:

$$\widetilde{g}(z) = \sum_{n=1}^{\infty} a_n 2^n \Gamma\left(n + \frac{1}{2}\right)(-i)^n J_n(iz)$$

$$= \frac{1}{\sqrt{\pi}} \sum_{n=1}^{\infty} a_n (-i)^n (iz)^n \int_{-1}^{1} \left(1 - t^2\right)^{n-\frac{1}{2}} \exp(-zt)dt. \tag{4.20}$$

Since $\left|(1 - t^2)\exp(-zt)\right| \leq e^R$ for $t \in [-1, 1]$, $n = 1, 2, \ldots$, the convergence of the expansion $\sum_{n=1}^{\infty} a_n z^n \left(1 - t^2\right)^{n-\frac{1}{2}} \exp(-zt)$ is uniform for t in the above interval. Because of that, $\widetilde{g}$ can be written in the following way:

$$\widetilde{g}(z) = \frac{1}{\sqrt{\pi}} \int_{-1}^{1} \left(\sum_{n=0}^{\infty} a_n \left(z\left(1 - t^2\right)\right)\right) \frac{\exp(-zt)}{\sqrt{1 - t^2}} dt.$$

From the inequality $\left|z\left(1 - t^2\right)\right| < R$ and equality (4.13), it follows that

$$\widetilde{g}(z) = \frac{1}{\sqrt{\pi}} \int_{-1}^{1} \frac{f\left(z\left(1 - t^2\right)\right)}{\sqrt{1 - t^2}} \exp(-zt)dt,$$

which is equivalent to (4.18). □

Theorem 4.5. *The main star S_f of the function $f(z)$ is contained in the main star S_g of the function $g(z)$, i.e. $S_f \subset S_g$.*

Proof. Let $z_0 \in S_f$. Then there exists a neighborhood U, starlike with respect to the origin and including the segment $[0, z_0]$, such that the function $f(z)$ is radially analytically continuable into U and $f(z)$ is continuous in the closed domain $[U] = U \cup \partial U$. Hence, there exists a constant $M > 0$ such that

$$\left| f\left(z\left(1 - t^2\right)\right) \cos(zt) \right| \leq \frac{M}{\sqrt{\pi}} \ , \quad z \in U.$$

Hence, we get the estimation

$$|g(z)| = \left| \frac{2}{\sqrt{\pi}} \int_0^1 \frac{f\left(z(1 - t^2)\right)}{\sqrt{1 - t^2}} \cos(zt) \, dt \right| \leq \frac{2M}{\pi} \int_0^1 \frac{dt}{\sqrt{1 - t^2}} = M$$

for the modulus of the integral in (4.16), i.e. the integral is uniformly convergent inside the neighborhood U and therefore, the function $g(z)$ is analytically continuable into the domain U. This means that $z_0 \in S_g$, i.e. $S_f \subset S_g$. □

Through analogy we can prove the following theorem.

Theorem 4.6. *The main star S_f of the function $f(z)$ is contained in the main star $S_{\widetilde{g}}$ of the function $\widetilde{g}(z)$, i.e. $S_f \subset S_{\widetilde{g}}$.*

Remark 4.2. Note that the integral dependence (4.16) is given in [Watson (1949, p. 579, 16.2)]. The author emphasizes that the integral dependence is as a result of Italian mathematician Salvatore Pincherle, who mentioned that it can be used for finding a relation between the singularities of f and g, however no proof of this is shown in [Watson (1949)].

4.6. Pringsheim's Type Theorem and Vivanti–Dienes Type Theorem

In this chapter, we prove the Pringsheim's type theorem and Vivanti–Dienes type theorem for the functions g and $\widetilde{g}$, given by (4.14) and (4.15), respectively. Let A_θ denote an arbitrary angular domain with a size $\theta < \pi$ and vertex at the point $z = 0$.

Theorem 4.7 (of Pringsheim type). *If the terms of the complex sequence $\{a_n\}_{n=0}^{\infty}$ are situated in the domain A_θ and $R < \infty$ is defined by (4.12), then $z_0 = R$ is a singular point of the function $\widetilde{g}(z)$.*

Proof. Using both relations (4.15) and (4.19), we deduce (4.20). subsequently, denoting for convenience $\widetilde{a}_n = a_n 2^n \Gamma\left(n + \frac{1}{2}\right)(-i)^n$, the equality (4.20) yields the form:

$$\widetilde{g}(z) = \sum_{n=0}^{\infty} \widetilde{a}_n J_n(iz).$$

Now $\widetilde{g}(z)$, being a holomorphic function in the disk $D(0; R)$, can be represented by a power series there, i.e.

$$\widetilde{g}(z) = \sum_{n=0}^{\infty} b_n (iz)^n. \tag{4.21}$$

Since the coefficients b_n and $\widetilde{a}_n$ are connected by the following relation [Erdélyi *et al.* (1953, 7.10 (4))]:

$$b_n = \frac{1}{\Gamma(n+1)2^n} \sum_{m=0}^{\left[\frac{n}{2}\right]} (-1)^m \binom{n}{m} \widetilde{a}_{n-2m}, \tag{4.22}$$

i.e.

$$b_n = \frac{(-i)^n}{\Gamma(n+1)} \sum_{m=0}^{\left[\frac{n}{2}\right]} \binom{n}{m} \frac{\Gamma\left(n - 2m + \frac{1}{2}\right)}{2^{2m}} a_{n-2m},$$

and all the numbers

$$\frac{1}{\Gamma(n+1)} \binom{n}{m} \frac{\Gamma\left(n - 2m + \frac{1}{2}\right)}{2^{2m}},$$

used in the last formula, are positive, then the coefficients $\widetilde{b}_n$, given below by the linear combinations

$$\widetilde{b}_n = b_n i^n = \frac{1}{\Gamma(n+1)} \sum_{m=0}^{\left[\frac{n}{2}\right]} \binom{n}{m} \frac{\Gamma\left(n - 2m + \frac{1}{2}\right)}{2^{2m}} a_{n-2m}, \tag{4.23}$$

are situated in the domain A_θ as well. Therefore, according to the Pringsheim's theorem for power series (Theorem 4.3), the point $z_0 = R$ is a singular point of the series $\sum_{n=0}^{\infty} b_n i^n z^n$, which is exactly the right-hand side of (4.21). Hence, the point $z_0 = R$ is a singular one of the function $\widetilde{g}(z)$. $\qquad \square$

Remark 4.3. The result that both series, representing the function $\widetilde{g}(z)$, have the same radius of convergence can also be obtained directly after establishing that

$$\left(\limsup_{n\to\infty}(|\widetilde{b}_n|)^{\frac{1}{n}}\right) = \left(\limsup_{n\to\infty}(|a_n|)^{\frac{1}{n}}\right) = R^{-1}.$$

Corollary 4.2. *If the terms of the sequence $\{a_n\}_{n=0}^{\infty}$ are such that the numbers $a_n i^n$ $(n = 0, 1, 2, \ldots)$ are situated in the angular domain A_θ and $R < \infty$, then $z_0 = iR$ is a singular point of the function $g(z)$.*

Proof. In fact, as a result of (4.14) and (4.19), we have

$$g(z) = \sum_{n=0}^{\infty} a_n 2^n \Gamma\left(n + \frac{1}{2}\right)(-i)^{-n} I_n(-iz),$$

and then, the substitution $\zeta = -iz$ leads to the series of the kind (4.15), containing $a_n i^n$ instead of a_n, namely

$$g(z) = g(i\zeta) = \sum_{n=0}^{\infty} a_n i^n 2^n \Gamma\left(n + \frac{1}{2}\right) I_n(\zeta). \tag{4.24}$$

Since $a_n i^n \in A_\theta$ $(n = 0, 1, 2, \ldots)$ and according to Theorem 4.7 the point $\zeta_0 = R$ is singular for the series (4.24), therefore, $z_0 = iR$ is a singular point of the function $g(z)$. $\qquad\square$

Theorem 4.8 (of Vivanti–Dienes type). *If the terms of the sequence $\{a_n\}_{n=0}^{\infty}$ satisfy the condition (4.12), and additionally*

$$a'_n = \Re(a_n) \geq 0 \quad (n = 0, 1, 2, \ldots)$$

and

$$\left(\limsup_{n\to\infty}(|a'_n|)^{\frac{1}{n}}\right)^{-1} = R < \infty,$$

then the point $z_0 = R$ is a singular one for the function $\widetilde{g}(z)$.

Proof. Through analogy with Theorem 4.7, we obtain equality (4.20) from the equality (4.15). Using the relation (4.22) between the coefficients of the series (4.20) and (4.21), we get (4.23). Now, in view of the notation $\widetilde{b}_n = b_n i^n$, introduced by (4.23), the series (4.21) can be written in the following way

$$\widetilde{g}(z) = \sum_{n=0}^{\infty} \widetilde{b}_n z^n, \quad |z| < R. \tag{4.25}$$

Note that the condition $a'_n = \Re(a_n) \geq 0$ implies the assertion $\widetilde{b}'_n = \Re(\widetilde{b}_n) \geq 0$. Moreover, in view of Remark 4.3, applied for a'_n (the radius is also R) and $\widetilde{b}'_n$ instead of a_n and $\widetilde{b}_n$, the equality

$$\left(\limsup_{n \to \infty} (|\widetilde{b}'_n|)^{\frac{1}{n}} \right)^{-1} = \left(\limsup_{n \to \infty} (|a'_n|)^{\frac{1}{n}} \right)^{-1} = R$$

holds. Now, from the classical Vivanti–Dienes theorem (Theorem 4.4), applied to the series (4.25), we conclude that $z_0 = R$ is a singular point of the function $\widetilde{g}(z)$. $\qquad \square$

Corollary 4.3. *If the terms of the sequence $\{a_n\}_{n=0}^{\infty}$ satisfy the condition (4.12), and additionally*

$$a'_n = \Re(i^n a_n) \geq 0 \quad (n = 0, 1, 2, \dots)$$

and

$$\left(\limsup_{n \to \infty} (|a'_n|)^{\frac{1}{n}} \right)^{-1} = R < \infty,$$

then the point $z_0 = iR$ is singular for the function $g(z)$.

Proof. Through analogy with the proof of Corollary 4.2, the function $g(z)$ can be written in the form (4.24). Its right-hand side has the kind (4.15), containing $i^n a_n$ instead of a_n. Further, bearing in mind that $\Re(i^n a_n) \geq 0$ and according to Theorem 4.8, we can conclude that $\zeta_0 = R$ is a singular point of its sum $g(i\zeta)$, from which $z_0 = iR$ is a singular point of the function $g(z)$. $\qquad \square$

Chapter 5

The Completeness of Systems of Bessel and Associated Bessel Functions in Spaces of Holomorphic Functions

5.1. The Completeness of Systems of Holomorphic Functions

Let G be an arbitrary region in the complex plane $\mathbb{C}$ and $H(G)$ be the space of the complex-valued functions holomorphic in G. As usual, we consider $H(G)$ with the topology of uniform convergence on the compact subsets of the domain G.

Definition 5.1. A system $\{\varphi_n(z)\}_{n=0}^{\infty} \subset H(G)$ is called *complete* in $H(G)$, if for every $f \in H(G)$, every compact set $K \subset G$ and every $\varepsilon > 0$, there exists a linear combination

$$P(z) = \sum_{n=0}^{N} c_n \varphi_n(z) \quad (c_n \in \mathbb{C}; \ n = 0, 1, 2, \ldots, N),$$

such that $|f(z) - P(z)| < \varepsilon$ whenever $z \in K$.

For example, if $G_1 \subset \mathbb{C}$ and $G_2 \subset \mathbb{C}\backslash\{0\}$ are simply connected regions, then the systems $\{z^n\}_{n=0}^{\infty}$ and $\{z^{-n}\}_{n=0}^{\infty}$ are complete in $H(G_1)$ and $H(G_2)$, respectively, and this assertion is nothing but a particular case of the Runge's approximation theorem [Zygmund and Saks (1952, p. 176, (2.1))].

Definition 5.2. Let γ be a Jordan curve (i.e. a simple closed curve) in $\mathbb{C}$ and C_γ be the closure of its outside with respect to the extended complex plane $\overline{\mathbb{C}} = \mathbb{C} \cup \{\infty\}$. Through H_γ, we denote the (vector) space of the complex-valued functions, each of which is holomorphic in an open set containing C_γ, and it vanishes at the infinite point.

The following statement is a criterion for completeness in a space of the kind $H(G)$ [Levin (1964, p. 211, Theorem 17)].

Criterion 5.1 (CC). A system $\{\varphi_n(z)\}_{n=0}^{\infty}$ of complex-valued functions, holomorphic in a simply connected region $G \subset \mathbb{C}$, is complete in the space $H(G)$ iff for every rectifiable Jordan curve $\gamma \subset G$ and every function $F \in H_\gamma$ the equalities:

$$\int_\gamma F(z)\varphi_n(z)dz = 0, \quad n = 0, 1, 2, \ldots,$$

imply $F = 0$.

Let $s = \{r_n\}_{n=0}^{\infty}$ be an increasing sequence of nonnegative real numbers, such that $\liminf_{n\to\infty}(r_{n+1} - r_n) > 0$. We denote by $N(s;r)$ $(0 \leq r < \infty)$ the number of these r_n which are not greater than r. Clearly, $N(s;r)$ is an increasing function and $N(s;r) = n + 1$ when $r_n \leq r < r_{n+1}$ $(n = 0, 1, 2, \ldots)$.

Definition 5.3. The limit

$$\delta(s) = \lim_{r\to\infty} (r^{-1}N(s;r)),$$

if it exists, is said to be a *density* of the sequence s.

Some well-known results concerning the density are listed below.

Assertion 5.1. The sequence $s = \{r_n\}_{n=0}^{\infty}$ has a density $\delta(s)$ iff there exists the limit

$$\delta(s) = \lim_{n\to\infty} \left(\frac{n}{r_n}\right).$$

Assertion 5.2. Let $0 < \xi < 1$. Then there exists the limit

$$\Delta(s) = \lim_{\xi\to1} \limsup_{r\to\infty} \left(\frac{N(s;r) - N(s;\xi r)}{r(1-\xi)}\right).$$

Definition 5.4. The value $\Delta(s)$ is said to be a *maximal density* of the sequence s.

It can be easily seen that if the series s has a density $\delta(s)$, then $\Delta(s) = \delta(s)$. So, if $0 < \xi < 1$, then

$$\Delta(s) = \lim_{\xi \to 1} \limsup_{r \to \infty} \left(\frac{N(s;r) - N(s;\xi r)}{r(1 - \xi)} \right)$$

$$= \lim_{\xi \to 1} \limsup_{r \to \infty} \left(\frac{N(s;r)}{r(1 - \xi)} - \frac{N(s;\xi r)}{r(1 - \xi)} \right)$$

$$= \lim_{\xi \to 1} \left(\frac{\delta(s)}{1 - \xi} - \frac{\delta(s)\xi}{1 - \xi} \right) = \delta(s).$$

Let $S^*(s)$ be the set of all the sequences that have densities and such that s is their subsequence, i.e.

$$S^*(s) = \{s^* : \exists\, \delta(s^*) \quad \text{and } s \text{ is a subsequence of } s^*\}.$$

Assertion 5.3. If $S^*(s)$ is a nonempty set, then

$$\Delta(s) = \inf \{\delta(s^*) : s^* \in S^*(s)\}.$$

Let $k = \{k_n\}_{n=0}^{\infty}$ be an increasing sequence of nonnegative integers, and $\tilde{k}$ be its complementary sequence, and suppose that $\tilde{k}$ is not finite.

Assertion 5.4. If $k = \{k_n\}_{n=0}^{\infty}$ has a positive density, then $\tilde{k}$ also has a density and

$$\delta(\tilde{k}) = 1 - \delta(k).$$

Proof. Therefore, since $N(k;k_n) + N(\tilde{k};k_n) = k_n + 1$ and $N(k;k_n) = n + 1$, we find that $N(\tilde{k};k_n) = k_n - n$. Let $k_n \leq r < k_{n+1}$, then $k_n - n \leq N(\tilde{k};r) \leq k_{n+1} - (n+1)$ and therefore,

$$\frac{k_n}{k_{n+1}} - \frac{n}{k_{n+1}} \leq \frac{N(\tilde{k};r)}{r} \leq \frac{k_{n+1}}{k_n} - \frac{n}{k_n} - \frac{1}{k_n}.$$

Since $\delta(k) > 0$, then $\lim_{n \to \infty} \frac{k_n}{k_{n+1}} = (\delta(k))^{-1}\delta(k) = 1$, from which we obtain that $\lim_{r \to \infty} \frac{N(\tilde{k};r)}{r} = 1 - \delta(k)$. $\qquad\square$

Remark 5.1. Note that

(i) Assertion 5.1 is well known [Pólya and Szego (1925, Vol. 1, Chapter 2, Problem 148)].
(ii) Assertion 5.2, as well as the notion 'maximal density' is due to Pólya [1929, p. 625, Theorem 7].
(iii) The proof of Assertion 5.3 can be found in [Bernstein (1933, Remark II.2)].
(iv) The given proof of Assertion 5.4 is explained in [Rusev (1997)].

Definition 5.5. Let $\sum_{n=0}^{\infty} a_n z^n$ be a power series with a positive and finite radius R of convergence. The point z_0 from the circle $C(0; R)$ is said to be a *singular point* for this power series if its sum cannot be continued analytically along the segment $[0, z_0]$.

According to a theorem of Eugène Fabry [Bieberbach (1955, Theorem 2.2.1)], if the density of the sequence indices of the coefficients (which are different from zero) of the considered power series is equal to zero, then each point on the circle $C(0; R)$ is a singular one, i.e. this circle is a natural boundary of the power series.

A remarkable generalization of this assertion is the following theorem of Pólya [1929, p. 625, Theorem IV.A].

Theorem 5.1 (of Pólya). *If the maximal density of the indices of the coefficients (which are different from zero) of a power series, with a radius of convergence R $(0 < R < \infty)$, is equal to Δ, then each closed arc from the boundary $C(0; R)$ of the disk of convergence, corresponding to a central angle $2\pi\Delta$, contains at least one singular point of the power series.*

Remark 5.2. Each point on $C(0; R)$ is considered as a closed arc corresponding to the angle 0.

Corollary 5.1. *Let $a_{k_n} = 0$ $(n = 0, 1, 2, \dots)$ and the sequence $k = \{k_n\}_{n=0}^{\infty}$ have a positive density $\delta(k)$. Then each arc on the circle $C(0; R)$ with an angle $2\pi(1 - \delta(k))$ contains at least one singular point of the power series $\sum_{n=0}^{\infty} a_n z^n$.*

Proof. Therefore, the sequence $\tilde{k}$, that is complementary sequence of the sequence k, has a density $\delta(\tilde{k}) = 1 - \delta(k)$ (according to Assertion 5.4) and moreover, $\Delta \leq \delta(\tilde{k})$ (confer with Assertion 5.3).

$\square$

Remark 5.3. Last corollary was also proved by Pólya in his paper [Pólya (1927)], published two years before [Pólya (1929)].

The completeness of systems of special functions in spaces of holomorphic functions has also been considered by other authors, for example, by Kazmin, Leont'ev and Rusev.

In Chapter 3 of his monograph [Leont'ev (1980)], Leont'ev studies the completeness of power, exponential, functions 'close' to exponents, as well as other systems of functions in various domains. Other important properties of polynomial and trigonometric families of functions as well as some of their selected applications in approximation theory and computer aided geometric design can also be seen, for example, in [Mastroianni and Milovanovic (2008); Milovanovic *et al.* (1994)]. In the paper [Kazmin (1960)], Kazmin considers the completeness of subsystems of Hermite and Laguerre polynomials. In a series of papers [Rusev (1977b, 1980, 1994a, 1994b, 1995, 1997)], Rusev studies the completeness of systems of Hermite, Jacobi and Laguerre polynomials, Kummer and Weber–Hermite functions in spaces of holomorphic functions. In this chapter, mainly based on the paper [Paneva-Konovska (2003)], the completeness of Bessel and associated Bessel functions systems is studied as well.

5.2. Auxiliary Statements about the Generating Function of Bessel Functions of the First Kind

Let $J_\nu(z)$ be the Bessel function of the first kind (1.2) with an index ν and Φ be the generating function, defined by (2.1).

Let $0 < \alpha < 1$ and $0 \leq \theta < 2\pi$. We denote with $A_{\alpha,\theta}$ the angular set

$$A_{\alpha,\theta} = \{z : z \in \mathbb{C}, |\arg(z\exp(-i\theta))| \leq \alpha\pi\}. \tag{5.1}$$

Lemma 5.1. *Let $0 < \alpha < 1$, $0 \le \theta < 2\pi$, $G \subset A_{\alpha,\theta}$ be a simply connected region, $\gamma \subset G$ be a rectifiable Jordan curve, $F \in H_\gamma$, and F be not identically zero. Let*

$$f(w) = \int_\gamma F(z)\Phi(z,w)dz. \qquad (5.2)$$

Then there exists a real number β, $0 < \beta < \alpha$, such that for $f(w)$ there are no singular points out of the set $A_{\beta,-\theta}$.

Proof. Since γ is a compact set, there exists β, $0 < \beta < \alpha$, such that $\gamma \subset A_{\beta,\theta}$ and $\gamma \cap \partial A_{\beta,\theta} \ne \varnothing$. The values of w for which $1 - zw(1-t^2) = 0$ are $w_\lambda = \lambda/z$, $\lambda \in [1,\infty)$. Let $z \in \gamma$. Then $z \in A_{\beta,\theta}$, that means $z \exp(-i\theta) \in A_{\beta,0}$ as well as $(z \exp(-i\theta))^{-1} \in A_{\beta,0}$. Bearing in mind the identity $z^{-1} = (z \exp(-i\theta))^{-1} \exp(-i\theta)$, we obtain $w_\lambda \exp(i\theta) = \lambda(z \exp(-i\theta))^{-1} \in A_{\beta,0}$, from which $w_\lambda \in A_{\beta,-\theta}$.

Hence, all the points w_λ for which $1 - zw(1 - t^2) = 0$ are in the set $A_{\beta,-\theta}$, i.e. the function $[1 - zw(1 - t^2)]^{-1}$ is holomorphic out of the set $A_{\beta,-\theta}$. Then the function (5.2) is also holomorphic for $w \in \text{Ext } A_{\beta,-\theta}$. $\square$

Remark 5.4. Note that $|w_\lambda| = \lambda/|z| \ge 1/|z|$. Further, if $\sup_{z \in \gamma} |z| = r$, then the function (5.2) is holomorphic for $|w| < 1/r$.

Lemma 5.2. *Let $\Re(\nu) > -1/2$, $R > 0$, K be a compact subset of the closed disk $[D(0;R)]$, i.e. $K \subset [D(0;R)]$, $F(z)$ be continuous in the set K and $|w| < 1/R$. Then the series*

$$\sum_{n=0}^{\infty} F(z)2^{n+\nu}\Gamma\left(n+\nu+\frac{1}{2}\right)z^{-\nu}J_{n+\nu}(z)w^n \qquad (5.3)$$

is absolutely and uniformly convergent in the set K.

Proof. Letting $z \in K$, a constant $M \ge 0$ exists, such that $|F(z)| \le M$ for all the values of $z \in K$. Since $|z| \le R$, applying the equality (2.3), and considering the absolute value of the general term of the series (5.3), we obtain the following upper estimate:

$$|u_n(z, w)| = |F(z)2^{n+\nu}\Gamma\left(n + \nu + \frac{1}{2}\right)z^{-\nu}J_{n+\nu}(z)w^n|$$

$$\leq M\pi^{-\frac{1}{2}}\left|\int_{-1}^{1}\exp(izt)(1 - t^2)^{n+\nu-\frac{1}{2}}(zw)^n dt\right|$$

$$\leq M\pi^{-\frac{1}{2}}|Rw|^n\exp(R)\int_{-1}^{1}(1 - t^2)^{\Re(\nu)-\frac{1}{2}}dt.$$

Now, since $\Re(\nu) - 1/2 > -1$, the integral

$$I = \int_{-1}^{1}(1 - t^2)^{\Re(\nu)-1/2}dt$$

converges and, denoting $S = MI\pi^{-1/2}\exp(R)$, we obtain that

$$|u_n(z, w)| \leq S|Rw|^n.$$

Finally, the convergence of the series

$$\sum_{n=0}^{\infty}|Rw|^n$$

leads to the absolute and uniform convergence of the considered series (5.3) in the compact set K. □

Theorem 5.2. *Let $0 < \alpha < 1$, $0 \leq \theta < 2\pi$, $G \subset A_{\alpha,\theta}$ be a simply connected region, $\gamma \subset G$ be a positive oriented rectifiable Jordan curve, $F \in H_\gamma$, F be not identically zero, and f be the function defined by (5.2). Then the expansion*

$$f(w) = \sum_{n=0}^{\infty}A_n(F)w^n \tag{5.4}$$

holds. The coefficients are

$$A_n(F) = 2^{n+\nu}\Gamma\left(n + \nu + \frac{1}{2}\right)\int_{\gamma}F(z)z^{-\nu}J_{n+\nu}(z)dz. \tag{5.5}$$

Moreover, the radius of convergence of the series (5.4) is finite.

Proof. Let $\max_{z \in \gamma} |z| = r$. Consider the complex-valued function (5.2) in the domain $|w| < 1/r$. Using Lemma 5.2, it is easy to obtain the expansion (5.4). Namely, in view of (2.1) and (5.2), we have

$$f(w) = \int_\gamma \left(F(z) \sum_{n=0}^{\infty} 2^{n+\nu} \Gamma\left(n + \nu + \frac{1}{2}\right) z^{-\nu} J_{n+\nu}(z) w^n \right) dz.$$

Then, according to the uniform convergence of the series (5.3), the positions of the integration and summation can be changed, i.e.

$$f(w) = \sum_{n=0}^{\infty} \left(2^{n+\nu} \Gamma\left(n + \nu + \frac{1}{2}\right) \int_\gamma F(z) z^{-\nu} J_{n+\nu}(z)\, dz \right) w^n,$$

and using the notation (5.5), we obtain the expansion (5.4), naturally with a radius of convergence $R \geq 1/r$. In fact,

$$R^{-1} = \limsup_{n \to \infty} |A_n(F)|^{\frac{1}{n}}.$$

It only remains to prove that $R \neq \infty$.

Let the curve Γ be a positive oriented circle with a center at 0 and a radius greater than r. Then the function $F(z) z^{-\nu} J_{n+\nu}(z)$ is holomorphic in the ring between γ and Γ. Therefore, the coefficients (5.5) can be represented in the form:

$$A_n(F) = 2^{n+\nu} \Gamma\left(n + \nu + \frac{1}{2}\right) \int_\Gamma F(z) z^{-\nu} J_{n+\nu}(z) dz. \qquad (5.6)$$

On the other hand, since $F(z)$ is holomorphic out of Γ, it can be represented there (see Theorem 4.2) by the series:

$$F(z) = \sum_{n=0}^{\infty} \tilde{A}_n(F) \tilde{A}_{n,\nu}(z), \qquad (5.7)$$

where

$$\tilde{A}_n(F) = (2\pi i)^{-1} \int_\Gamma F(\zeta) \zeta^{-\nu} J_{n+\nu}(\zeta) d\zeta \qquad (5.8)$$

and $\tilde{A}_{n,\nu}$ are the generalized Neumann polynomials (4.2).

According to Theorem 4.1, the radius of convergence $\tilde{R}$ of the series (5.7) is $\tilde{R} = 2 \limsup_{n \to \infty} |\tilde{A}_n(F) \Gamma(n + \nu + 1)|^{1/n}$. From (5.6)

and (5.8), it follows $\tilde{A}_n(F) = A_n(F)\,(2\pi i\,2^{n+\nu}\Gamma(n+\nu+1/2))^{-1}$.
Bearing in mind the Stirling's formula, the equality $\tilde{R} = 1/R$ can be deduced.

Let us suppose that $R = \infty$. Therefore, $\tilde{R} = 0$. It means (see Theorem 4.2) that $F(z)$ can be expanded in a convergent series in the functions $\tilde{A}_{n,\nu}(z)$ in the region $D^\star(0;0)$, given by (4.4). On the other hand, F is holomorphic on γ and in its outside, and hence also in the neighborhood of the point 0. Therefore, F is holomorphic in $\overline{\mathbb{C}}$ and $F(\infty) = 0$, from which it follows $F = 0$. Consequently, $R \neq \infty$. $\qquad\square$

5.3. Theorems about the Completeness for Bessel Functions Systems of the First Kind

The results, contained in this chapter, refer to the completeness of the functions system $\{z^{-\nu}J_{n+\nu}(z)\}_{n=0}^{\infty}$ and some of its subsystems in spaces of holomorphic functions. Sufficient conditions for such a type of completeness are given.

Theorem 5.3. *Let $G \subset \mathbb{C}$ be a simply connected region.*

(i) *If $-\nu \notin \mathbb{N}$, then the system $\{z^{-\nu}J_{n+\nu}(z)\}_{n=0}^{\infty}$ is complete in the space $H(G)$.*

(ii) *If $-\nu \in \mathbb{N}$ and $0 \notin G$, then the system $\{z^{-\nu}J_{n+\nu}(z)\}_{n=0}^{\infty}$ is complete in the space $H(G)$.*

Proof. Let $f(z)$ be a holomorphic function in G, K be a compact subset of G and $\varepsilon > 0$.

(i) Since the system $\{z^n\}_{n=0}^{\infty}$ is complete in $H(G)$, there exists a polynomial $P(z)$ such that $|f(z) - P(z)| < \varepsilon/2$, whenever $z \in K$.

Let $\nu \neq -1, -2, -3 \ldots$. The polynomial $P(z)$ is a holomorphic function in the complex plane and therefore, it can be represented in the form $P(z) = z^{-\nu}\sum_{n=0}^{\infty}a_n J_{n+\nu}(z)$, $a_n \in \mathbb{C}$ (see Theorem 4.2). The convergence of the series on the right-hand side is uniform in the set K. Consequently, there exists a natural number N such that $|P(z) - \sum_{n=0}^{N}a_n z^{-\nu}J_{n+\nu}(z)| < \varepsilon/2$, for each $z \in K$. Then we conclude that $|f(z) - \sum_{n=0}^{N}a_n z^{-\nu}J_{n+\nu}(z)| < \varepsilon$, from which the completeness of the system $\{z^{-\nu}J_{n+\nu}(z)\}_{n=0}^{\infty}$ in the space $H(G)$ follows.

Remark 5.5. Note that, in particular, $\{J_n(z)\}_{n=0}^{\infty}$ is complete in $H(G)$.

(ii) Let now $-\nu$ be a natural number and $0 \notin G$. Then, taking into account the holomorphicity of the function $z^{\nu}f(z)$ in the domain G, as well as the completeness of $\{J_n(z)\}_{n=0}^{\infty}$, the system $\{z^{-\nu}J_n(z)\}_{n=0}^{\infty}$ is also complete in $H(G)$. Since $\{z^{-\nu}J_n(z)\}_{n=0}^{\infty} \subset \{z^{-\nu}J_{n+\nu}(z)\}_{n=0}^{\infty}$, the system $\{z^{-\nu}J_{n+\nu}(z)\}_{n=0}^{\infty}$ is complete in $H(G)$, too. □

Remark 5.6. Let us emphasize that in the proof of Theorem 5.3, the fact that the system $\{z^{-\nu}J_{n+\nu}(z)\}_{n=0}^{\infty}$ is a basis in each space of the kind $H(D(0;R))$ $(0 < R \le \infty)$ is substantially exploited.

Theorem 5.4. *Let* $0 < \alpha < 1$, $0 \le \theta < 2\pi$, $\lim_{n\to\infty}(n/k_n) = \delta \ge \alpha$. *Then the system*

$$\left\{z^{-\nu}J_{k_n+\nu}(z)\right\}_{n=0}^{\infty} \tag{5.9}$$

is complete in the space $H(G)$ for every simply connected region $G \subset A_{\alpha,\theta}$, defined by (5.1).

Proof. First, let $\Re(\nu) > -1/2$. Suppose that the above assertion is not true. Then there exists a simply connected region $G \subset A_{\alpha,\theta}$, such that the system (5.9) is not complete in $H(G)$. In accordance with Criterion 5.1, there exist a rectifiable Jordan curve $\gamma \subset G$ and a function $F \in H_\gamma$, not identically zero, such that:

$$\int_\gamma F(z)z^{-\nu}J_{k_n+\nu}(z)dz = 0, \quad n = 0, 1, 2, \ldots. \tag{5.10}$$

Let us note that the function (5.2) is not identically zero, else, because of of the completeness of the whole system $\{z^{-\nu}J_{n+\nu}(z)\}_{n=0}^{\infty}$, it should be deduced that $F = 0$. Moreover, by using (5.4), (5.5) and (5.10), we obtain

$$f(w) = \sum_{n=0}^{\infty} a_{\tilde{k}_n}w^{\tilde{k}_n}, \quad \lim_{n\to\infty}\left(\frac{n}{\tilde{k}_n}\right) = 1 - \delta. \tag{5.11}$$

Further, let Δ be the maximal density of the increasing sequence of indices of nonzero coefficients in the series (5.11). This sequence is a subsequence of $\{\widetilde{k}_n\} = \{n\}\backslash\{k_n\}$. Then according to the statements in Sec. 5.1, related to the maximal density, from $0 < \alpha < 1$ and $\delta \geq \alpha$, it follows that

$$\Delta \leq 1 - \delta \leq 1 - \alpha. \tag{5.12}$$

Since $F(z)$ is not identically zero, not all of the complex numbers (5.5) are equal to zero. Then according to Lemma 5.1, there exists a number β, $0 < \beta < \alpha$, such that all singular points of $f(w)$ are situated in the set $A_{\beta,-\theta}$. Let R be the radius of convergence of the series (5.4). According to Theorem 5.2, $0 < R < \infty$. Therefore, there exists a closed arc of the circle $|w| = R$ with length $2\pi(1-\beta)R$, where (5.4) has no singular points.

On the other hand, by Theorem 5.1 of George Pólya, every closed arc of $|w| = R$ with length $2\pi\Delta R$ contains at least one singular point of $f(w)$. Because of (5.12), $2\pi\Delta \leq 2\pi(1-\delta) \leq 2\pi(1-\alpha) < 2\pi(1-\beta)$ and we come to a contradiction.

Hence, the system (5.9) is complete in every space $H(G)$, provided that $G \subset A_{\alpha,\theta}$ is simply connected.

Now, let $\Re(\nu) \leq -1/2$ and $z \in G$. Consider again the system (5.9). Let k_p be the minimal index k_n for which $\Re(k_n + \nu) > -1/2$. Then we have:

$$z^{-k_p-\nu} J_{k_n+\nu}(z) = z^{-k_p-\nu} J_{k_p+\nu+k_n-k_p}(z),$$

$$\Re(k_p + \nu) > -\frac{1}{2}, \quad \lim_{n\to\infty}\left(\frac{n}{(k_n - k_p)}\right) = \delta.$$

Hence, according to the proof above, the system $\{z^{-k_p-\nu} J_{k_n+\nu}(z)\}_{n=p}^{\infty}$ is complete in $H(G)$. Taking into account that

$$\left\{z^{-k_p-\nu} J_{k_n+\nu}(z)\right\}_{n=p}^{\infty} \subset \left\{z^{-k_p-\nu} J_{k_n+\nu}(z)\right\}_{n=0}^{\infty},$$

the completeness of the last system follows immediately. Since $0 \notin G$, then the system $\{z^{k_p} z^{-k_p-\nu} J_{k_n+\nu}(z)\}_{n=0}^{\infty} = \{z^{-\nu} J_{k_n+\nu}(z)\}_{n=0}^{\infty}$ is complete in $H(G)$, too. $\qquad\square$

5.4. Auxiliary Statements about the Generating Function of Modified Bessel Functions of Third Kind

Let Φ be the function defined by (2.16). Let $0 < \alpha < 1$ and A_α be the set defined by (2.5). The following lemma holds.

Lemma 5.3. *Let $G \subset A_\alpha$ $(0 < \alpha < 1)$ be a simply connected region, $\gamma \subset G$ be a rectifiable Jordan curve, $F \in H_\gamma$, and F be not identically zero. Let*

$$f(w) = \int_\gamma F(z)\Phi(z, w)dz. \tag{5.13}$$

Then there exists a real number β, $0 < \beta < \alpha$, such that $f(w)$ has no singular points outside the set A_β.

Proof. Since γ is a compact set, there exists a set A_β $(0 < \beta < \alpha)$ of the kind (2.5) such that $\gamma \subset A_\beta$ and $\gamma \cap \partial A_\beta \neq \varnothing$, i.e. at least one arm of the angle A_β touches the curve range. The values of w, for which

$$z^2 + t^2 - zw = 0, \tag{5.14}$$

are $w_t = z + z^{-1}t^2$ with $t \in [0, \infty)$. Let $z \in \gamma$, then $z^{-1} \in A_\beta$ and $w_t \in A_\beta$, as well. Therefore, all the points for which (5.14) holds are in A_β and outside the set A_β, the function $(z^2 + t^2 - zw)^{-1}$ is a holomorphic function of w. Therefore, the function (5.13) is holomorphic for the values of $w \in \text{Ext}A_\beta$, too. $\qquad\square$

Let $K_{n+1/2}$ $(n = 0, 1, 2, \dots)$ be the modified Bessel functions (1.9) of the third kind with indices $n + 1/2$, i.e.

$$K_{n+\frac{1}{2}}(z) = \sqrt{\frac{\pi}{2z}} \exp(-z) \sum_{k=0}^{n} \frac{(2z)^{-k}\Gamma(n+k+1)}{k!\,\Gamma(n-k+1)}, \quad |\arg z| < \pi.$$

Lemma 5.4. *Let K be a compact subset of the half-plane $\{z : z \in \mathbb{C}, \Re(z) > 0\}$, $\inf_{z \in K} |z| = r$, ρ_0 be defined by (2.7), $z \in K$, F be* *continuous function in K, and $|w| < \rho_0$. Then the series*

$$\sum_{n=0}^{\infty} F(z) \frac{K_{n+\frac{1}{2}}(z)}{2^{n+\frac{1}{2}}\,\Gamma(n+1)}\, w^n \tag{5.15}$$

is absolutely and uniformly convergent in the set K.

Proof. Since K is a compact set, there exists a positive constant $R > 0$, such that $\sup_{z \in K} |z| = R$. Denoting by $u_n(z, w)$ the general term of the series (5.15), in accordance with the integral representation (1.18), we have

$$u_n(z, w) = \sqrt{\frac{z}{\pi}}\, F(z) \int_0^\infty \frac{\cos t}{z^2 + t^2} \left(\frac{zw}{z^2 + t^2} \right)^n dt$$

$$= \sqrt{\frac{z}{\pi}}\, F(z) \int_0^{R+1} \frac{\cos t}{z^2 + t^2} \left(\frac{zw}{z^2 + t^2} \right)^n dt$$

$$+ \sqrt{\frac{z}{\pi}}\, F(z) \int_{R+1}^\infty \frac{\cos t}{z^2 + t^2} \left(\frac{zw}{z^2 + t^2} \right)^n dt.$$

Setting

$$M_0 = \max \left| \sqrt{\frac{z}{\pi}}\, \frac{F(z)}{z^2 + t^2} \right|, \quad z \in K, \ t \in [0, R+1],$$

$$M_1 = \max \left| \sqrt{\frac{z}{\pi}}\, F(z) \right|, \quad z \in K,$$

$$M = M_0(R+1) + M_1 \frac{\ln(2R+1)}{2R},$$

we can write the following estimation for the absolute value of $u_n(z, w)$:

$$|u_n(z, w)| \leq M \left(\frac{|w|}{\rho_0} \right)^n.$$

Now, keeping in view the convergence of the power series $\sum_{n=0}^{\infty} \left(\frac{|w|}{\rho_0} \right)^n$, the series (5.15) is absolutely and uniformly convergent on the compact set K. $\qquad \square$

Theorem 5.5. *Let $G \subset A_\alpha$ ($0 < \alpha < 1$) be a simply connected region, $\gamma \subset G$ be a rectifiable Jordan curve, $\inf_{z \in \gamma} |z| = r$, ρ_0 be defined by (2.7), $F \in H_\gamma$, F be not identically zero, $|w| < \rho_0$ and $f(w)$ be the function, defined by (5.13). Then the expansion*

$$f(w) = \sum_{n=0}^{\infty} A_n(F)w^n \qquad (5.16)$$

holds, with coefficients

$$A_n(F) = \frac{1}{2^{n+\frac{1}{2}} \, \Gamma(n+1)} \int_\gamma F(z) K_{n+\frac{1}{2}}(z) dz. \qquad (5.17)$$

Moreover, the series in (5.16) has a finite radius of convergence.

Proof. Considering the complex function $f(w)$, we have

$$f(w) = \int_\gamma \left(F(z) \sum_{n=0}^{\infty} \frac{K_{n+\frac{1}{2}}(z)}{2^{n+\frac{1}{2}} \, \Gamma(n+1)} \, w^n \right) dz$$

$$= \sum_{n=0}^{\infty} \left(\frac{1}{2^{n+\frac{1}{2}} \, \Gamma(n+1)} \int_\gamma F(z) K_{n+\frac{1}{2}}(z) dz \right) w^n, \qquad (5.18)$$

and then, using the denotation (5.17), we obtain (5.16).

Since the curve γ is a compact set, $F \in H_\gamma$, and according to Lemma 5.4, the uniform convergence of the series (5.15) follows immediately. That is why the change of the positions of integration and summing in (5.18) is legal.

Suppose that the series in (5.16) has an infinite radius of convergence. This means that the equality (5.16) represents an entire function, i.e. $f(w)$ is an entire function. In order to determine the order of $f(w)$, we find an upper estimation of $|f(w)|$. From the equalities (5.13) and (2.19), we obtain consequently:

$$|f(w)| = \frac{\sqrt{\pi}}{2} \left| \int_\gamma \frac{F(z) \exp(-\sqrt{z}\sqrt{z-w})}{\sqrt{z-w}} \, dz \right|$$

$$\leq \frac{\sqrt{\pi}}{2} \int_\gamma \frac{|F(z)| \exp(\sqrt{|z|}\sqrt{|z-w|})}{\sqrt{|z-w|}} \, ds$$

$$= \frac{\sqrt{\pi}}{2} \int_\gamma \frac{|F(z)| \exp(\sqrt{|z|}\sqrt{|w|}\sqrt{|1 - zw^{-1}|})}{\sqrt{|z-w|}} \, ds.$$

Since $\lim_{|w|\to\infty}(1 - zw^{-1}) = 1$ and $\lim_{|w|\to\infty}(z - w)^{-1/2} = 0$, then for sufficiently large $|w|$, the inequalities $|1 - zw^{-1}|^{1/2} < 2$ and $|z - w|^{-1/2} < 2$ hold. Denoting

$$m = \sup_{z\in\gamma}|F(z)|, \quad L = l(\gamma) = \text{length}(\gamma),$$

$$M = \sqrt{\pi}\,mL, \quad R = \sup_{z\in\gamma}|z|,$$

we obtain that there exists a constant $B > 0$, such that for all the values of $|w| > B$, the inequality

$$|f(w)| \le \frac{\sqrt{\pi}}{2}\,2mL\,\exp\left(2\sqrt{R}|w|^{\frac{1}{2}}\right) = M\exp\left(2\sqrt{R}|w|^{\frac{1}{2}}\right)$$

holds, i.e.

$$|f(w)| \le M\exp\left(2\sqrt{R}|w|^{\frac{1}{2}}\right), \quad |w| > B. \tag{5.19}$$

Therefore, the function $f(w)$ is of order $\rho \le 1/2$.

Furthermore, considering $f(iv)$ ($v \in (-\infty, \infty)$), we have

$$|f(iv)| = \frac{\sqrt{\pi}}{2}\left|\int_\gamma \frac{F(z)\exp(-\sqrt{z}\sqrt{z - iv}\,)}{\sqrt{z - iv}}\,dz\right|.$$

Since $\Re(z) > 0$, then $|\arg z| < \pi/2$ and $|\arg(z - iv)| < \pi/2$, i.e. denoting $z = r_1\exp(i\vartheta_1)$ and $z - iv = r_2\exp(i\vartheta_2)$, with $|\vartheta_{1,2}| < \pi/2$, we obtain

$$\exp(-\sqrt{z}\sqrt{z - iv}\,)$$
$$= \exp\left(-\sqrt{r_1 r_2}\exp\left(i\frac{\vartheta_1 + \vartheta_2}{2}\right)\right)$$
$$= \exp\left(-\sqrt{r_1 r_2}\cos\left(\frac{\vartheta_1 + \vartheta_2}{2}\right) - i\sqrt{r_1 r_2}\sin\left(\frac{\vartheta_1 + \vartheta_2}{2}\right)\right),$$

from which

$$|\exp(-\sqrt{z}\sqrt{z - iv}\,)| = \exp\left(-\sqrt{r_1 r_2}\cos\left(\frac{\vartheta_1 + \vartheta_2}{2}\right)\right),$$

with $\left|\frac{\vartheta_1 + \vartheta_2}{2}\right| < \frac{\pi}{2}$, and hence $\cos\left(\frac{\vartheta_1 + \vartheta_2}{2}\right) > 0$.

Let $z = x + iy$, then $z - iv$ satisfies the equality $|z - iv| = \sqrt{x^2 + (y - v)^2}$, from which $|z - iv|^{-1/2} \leq x^{-1/2}$. Then we have

$$|f(iv)| \leq \frac{\sqrt{\pi}}{2} \int_\gamma \frac{|F(z)| \, |\exp(-\sqrt{z}\sqrt{z - iv})|}{\sqrt{|z - iv|}} \, ds. \qquad (5.20)$$

Denoting

$$m_1 = \sup_{z \in \gamma} |F(z)|, \quad m_2 = \sup_{z \in \gamma} x^{-\frac{1}{2}},$$

$$L = l(\gamma), \quad M = \sqrt{\pi} \, m_1 m_2 \frac{L}{2}, \qquad (5.21)$$

we obtain consequently:

$$|f(iv)| \leq \frac{\sqrt{\pi}}{2} m_1 m_2 L \exp\left(-\sqrt{r_1 r_2} \cos\left(\frac{\vartheta_1 + \vartheta_2}{2}\right)\right)$$

$$\leq M, \quad v \in (-\infty, \infty). \qquad (5.22)$$

From (5.19) and (5.22), according to the Phragmén–Lindelöf theorem, it follows that $|f(w)| \leq M$ for every $w \in \mathbb{C}$. In accordance with the above made supposition, $f(w)$ is an entire function, and therefore, $f(w) = const.$

Considering (5.20) for sufficiently large v, we have

$$|f(iv)| \leq m_1 \frac{\sqrt{\pi}}{2} \int_\gamma \frac{\left|\exp\left(-\sqrt{z}\sqrt{z - iv}\right)\right|}{\sqrt{|v|}\sqrt{|1 - z(iv)^{-1}|}} \, ds.$$

With regard to the denotation (5.21) and inequality $|1 - z(iv)^{-1}| > 1/2$, we obtain that, for sufficiently large v, the inequality $|f(iv)| \leq \frac{\sqrt{\pi}}{2} 2m_1 L |v|^{-1/2}$ holds. Now, since $\lim_{n \to \infty} |v|^{-1/2} = 0$, it follows that $f(w) \equiv 0$. The last equality leads to contradiction with the Criterion 5.1 for completeness, since F is not identically zero and the system $\{K_{n+1/2}(z)\}_{n=0}^{\infty}$ is complete in $H(G)$ (Theorem 5.6). Therefore, the series (5.16) has a finite radius of convergence. $\square$

5.5. Theorems about the Completeness for Modified Bessel Functions Systems of Third Kind

The assertions obtained in this chapter are related to the completeness of families of modified Bessel functions of the third kind and some of their subsystems.

Theorem 5.6. *Let* $G \subset \mathbb{C} \backslash (-\infty, 0]$ *be a simply connected region. Then the system* $\left\{ K_{n+1/2}(z) \right\}_{n=0}^{\infty}$ *is complete in the space* $H(G)$.

Proof. Let $\gamma \subset G$ be a rectifiable Jordan curve and $F \in H_\gamma$. Let

$$\int_\gamma F(z) K_{n+\frac{1}{2}}(z) dz = 0, \quad n = 0, 1, 2, \ldots. \tag{5.23}$$

By induction it can be proved that

$$\int_\gamma F(z) \sqrt{\frac{\pi}{2z}} \, z^{-n} \exp(-z) dz = 0, \quad n = 0, 1, 2, \ldots. \tag{5.24}$$

So, let first $n = 0$. Then, in view of (1.10), we have that $K_{1/2}(z) = \sqrt{\frac{\pi}{2z}} \exp(-z)$ and therefore, according to (5.23),

$$\int_\gamma F(z) K_{\frac{1}{2}}(z) dz = \int_\gamma F(z) \sqrt{\frac{\pi}{2z}} \exp(-z) dz = 0. \tag{5.25}$$

For $n = 1$, we obtain $K_{3/2}(z) = \sqrt{\frac{\pi}{2z}} \left(1 + z^{-1}\right) \exp(-z)$ and subsequently, with the aid of (5.23) and (5.25), we have

$$0 \neq \int_\gamma F(z) K_{\frac{3}{2}}(z) dz$$

$$= \int_\gamma F(z) \sqrt{\frac{\pi}{2z}} \exp(-z) dz + \int_\gamma F(z) \sqrt{\frac{\pi}{2z}} \, z^{-1} \exp(-z) dz$$

$$= \int_\gamma F(z) \sqrt{\frac{\pi}{2z}} \, z^{-1} \exp(-z) dz.$$

Supposing that

$$\int_\gamma F(z) \sqrt{\frac{\pi}{2z}} \, z^{-n} \exp(-z) dz = 0, \quad n = 0, 1, \ldots, s, \tag{5.26}$$

we consider $\int_\gamma F(z) K_{s+3/2}(z) dz$. According to (1.9), (5.23) and (5.26), we obtain:

$$0 = \int_\gamma \left(F(z) \sqrt{\frac{\pi}{2z}} \exp(-z) \sum_{m=0}^{s+1} (2z)^{-m} \frac{\Gamma(s+1+m+1)}{m! \Gamma(s+1-m+1)} \right) dz$$

$$= 2^{-s-1} \frac{\Gamma(2s+3)}{(s+1)! \Gamma(1)} \int_\gamma \left(F(z) \sqrt{\frac{\pi}{2z}} z^{-s-1} \exp(-z) \right) dz,$$

that proves (5.24).

Further, since the system $\{z^{-n}\}_{n=0}^\infty$ is complete in the space $H(G)$ and $\sqrt{\frac{\pi}{2z}} \exp(-z) \neq 0$ for $z \in G$, then the system $\left\{ \sqrt{\frac{\pi}{2z}} z^{-n} \exp(-z) \right\}_{n=0}^\infty$ is complete in $H(G)$, too. Then (5.24) and Criterion 5.1 lead to the identity $F = 0$, from which the completeness of the system $\left\{ K_{n+1/2}(z) \right\}_{n=0}^\infty$ follows. $\qquad\square$

Theorem 5.7. *Let $0 < \alpha < 1$ and $\lim_{n\to\infty}(n/m_n) = \delta \geq \alpha/2$. Then the system*

$$\left\{ K_{m_n + \frac{1}{2}}(z) \right\}_{n=0}^\infty \tag{5.27}$$

is complete in the space $H(G)$ for every simply connected region $G \subset A_\alpha$ of the kind (2.5).

Proof. Let us suppose that $\lim_{n\to\infty}(n/m_n) = \delta \geq \alpha/2$, but nevertheless there exists a simply connected region $G \subset A_\alpha$, such that the family (5.27) is not complete in the space $H(G)$. This means that there exist a rectifiable Jordan curve $\gamma \subset G$ and a function $F \in H_\gamma$, F is not identically zero but

$$\int_\gamma F(z) K_{m_n + \frac{1}{2}}(z) dz = 0, \quad n = 0, 1, 2, \dots. \tag{5.28}$$

Let $r = \inf_{z\in\gamma} |z|$ and ρ_0 be defined by (2.7). Consider the complex-valued function $f(w)$ defined by (5.13) on the circular domain $|w| < \rho_0$. Let us note that the function (5.13) is not identically zero (the opposite one, because of the completeness of the whole family of functions, would mean that $F = 0$). Moreover, by using (5.16), (5.17) and (5.28), we obtain

$$f(w) = \sum_{n=0}^{\infty} A_{\widetilde{m}_n}(F) w^{\widetilde{m}_n}, \quad \widetilde{\delta} = \lim_{n \to \infty} \left(\frac{n}{\widetilde{m}_n} \right) = 1 - \delta. \qquad (5.29)$$

Furthermore, if Δ is the maximal density of the increasing sequence of indices of the nonzero coefficients in the series (5.29), according to the statements in Sec. 5.1 related to the maximal density, from $0 < \alpha < 1$ and $\delta \geq \alpha/2$, it follows that

$$\Delta \leq \widetilde{\delta} = 1 - \delta \leq 1 - \frac{\alpha}{2}. \qquad (5.30)$$

Since F is not identically zero, not all of the complex numbers (5.17) are equal to zero. Then according to Lemma 5.3, there exists a number β, $0 < \beta < \alpha$, such that all the singular points of $f(w)$ on the circle $|w| = R$ (R is the radius of convergence of (5.16)) are situated in the set A_β, i.e. there exists a closed arc corresponding to a central angle with a size of $\pi(2 - \beta)$, where (5.16) has no singular points. On the other hand, according to the Pólya Theorem 5.1, every closed arc of the circle $|w| = R$ with corresponding central angle $2\pi\Delta$ contains at least one singular point of (5.29). Following (5.30), we have

$$2\pi\Delta \leq 2\pi(1 - \delta) \leq 2\pi \left(1 - \frac{\alpha}{2} \right) < 2\pi \left(1 - \frac{\beta}{2} \right) = \pi(2 - \beta),$$

and this leads us to a contradiction. Hence, the system (5.27) is complete in every space $H(G)$, provided that the region $G \subset A_\alpha$ is simply connected. □

Corollary 5.2. *Let* $0 < \alpha < 1$ *and* $\lim_{n \to \infty}(n/m_n) = \delta \geq \alpha/2$. *Then the system*

$$\left\{ \sum_{k=0}^{m_n} 2^{-k} \frac{\Gamma(m_n + k + 1)}{k!\Gamma(m_n - k + 1)} z^k \right\}_{n=0}^{\infty} \qquad (5.31)$$

is complete in the space $H(G)$ *for every simply connected region* $G \subset A_\alpha$.

Proof. Since the system (5.27) is complete in the space $H(G)$ and $\sqrt{\frac{\pi}{2z}}\exp(-z) \neq 0$ for $z \in G$, then the system

$$\left\{\sum_{k=0}^{m_n} 2^{-k}\frac{\Gamma(m_n+k+1)}{k!\Gamma(m_n-k+1)}z^{-k}\right\}_{n=0}^{\infty}$$

is also complete in $H(G)$. Now, the completeness of (5.31) is obtained after changing of z with z^{-1}.   □

5.6. Auxiliary Statements about the Generating Function of Neumann Polynomials

Let Φ be the function defined by the relation (2.20).

Lemma 5.5. *Let G be a simply connected region in the right half-plane $\{z; z \in \mathbb{C}, \Re(z) > 0\}$, $\gamma \subset G$ be a rectifiable Jordan curve, $F \in H_\gamma$, and F be not identically zero. Let $\inf_{z \in \gamma} \Re(z) = \rho$, $|\Re(w)| < \rho/2$, and*

$$f(w) = \int_\gamma F(z)\Phi(z,w)dz. \tag{5.32}$$

Then the function (5.32) can be represented by the power series

$$f(w) = \sum_{n=0}^{\infty} A_n(F)w^n, \quad |w| < \frac{\rho}{2}, \tag{5.33}$$

with coefficients

$$A_n(F) = \frac{(-1)^n}{n!}\int_\gamma F(z)O_n(z)dz. \tag{5.34}$$

Proof. Since γ is a compact set in the right half-plane, then $\rho > 0$. Therefore, the inequalities $\Re(z) > 0$ and $\Re(z + 2w) > 0$ hold, that yield to the holomorphicity of both functions $\Phi(z,w)$ and $f(w)$ for $|\Re(w)| < \rho/2$. This shows that the function $f(w)$ can be represented by Taylor's series in the disk $|w| < \rho/2$ with coefficients $A_n(F) = \frac{f^{(n)}(0)}{n!}$. Bearing in mind the explicit form of the function

$f(w)$, we find

$$f^{(n)}(w) = \int_\gamma F(z)\Phi_w^{(n)}(z,w)dz.$$

Now, using (1.11) and (2.20), we obtain consecutively

$$2(-1)^n\Phi_w^{(n)}(z,w)$$

$$= \int_0^\infty \exp(-zt)(t+\sqrt{1+t^2})^n \exp(-w(t+\sqrt{1+t^2}))dt$$

$$+ \int_0^\infty \exp(-zt)(t-\sqrt{1+t^2})^n \exp(-w(t-\sqrt{1+t^2}))dt,$$

$$2(-1)^n\Phi_w^{(n)}(z,0)$$

$$= \int_0^\infty \exp(-zt)((t+\sqrt{1+t^2})^n + (t-\sqrt{1+t^2})^n)dt,$$

$$\Phi_w^{(n)}(z,0) = (-1)^n O_n(z), \quad f^{(n)}(0) = (-1)^n \int_\gamma F(z)O_n(z)dz,$$

from which the power expansion (5.33) is obtained with coefficients (5.34). □

Lemma 5.6. *Let G be a simply connected domain in the right half-plane $\{z : z \in \mathbb{C}, \Re(z) > 0\}$, $\gamma \subset G$ be a rectifiable Jordan curve, $F \in H_\gamma$, F be not identically zero and the function $f(w)$ be defined by (5.32). Then the series (5.33) with the coefficients (5.34) has a finite radius of convergence.*

Proof. We need to draw a circle C, having sufficiently small radius ($r_C < \inf_{z \in \gamma} |z|$) around the origin, and a circle Γ, having sufficiently large radius ($r_\Gamma > \sup_{z \in \gamma} |z|$). Orienting the curves γ, Γ and C in positive direction and integrating along Γ, we can write

$$\int_\Gamma F(z)O_n(z)dz = \int_\gamma F(z)O_n(z)dz + \int_C F(z)O_n(z)dz. \quad (5.35)$$

Since $F(\infty) = 0$ and $O_n(\infty) = 0$, the left-hand side of (5.35) is equal to zero. Therefore,

$$\int_\gamma F(z)O_n(z)dz = - \int_C F(z)O_n(z)dz,$$

from which (5.34) takes the form:

$$A_n(F) = \frac{(-1)^{n+1}}{n!} \int_C F(z)O_n(z)dz. \tag{5.36}$$

So, on the one hand, the function $f(w)$ can be expanded by the power series (5.33) with the coefficients (5.36) and a radius of convergence R,

$$R = \left(\limsup_{n \to \infty} |A_n(F)|^{\frac{1}{n}} \right)^{-1}.$$

On the other hand, the function F can be expanded by the Bessel function series in the disk, bounded by the circle C (in accordance with Theorem 4.2),

$$F(z) = \sum_{n=0}^{\infty} \widetilde{A}_n(F)J_n(z)dz,$$

with the coefficients

$$\widetilde{A}_n(F) = \frac{\varepsilon_n}{2\pi i} \int_C F(z)O_n(z)dz \tag{5.37}$$

and a radius of convergence (in accordance with Theorem 3.4)

$$\widetilde{R} = \left(\limsup_{n \to \infty} \left(|\widetilde{A}_n(F)| \, 2^{-n}(n!)^{-1} \right)^{\frac{1}{n}} \right)^{-1}.$$

Further, the following relationship

$$\widetilde{A}_n(F) = (-1)^{n+1} \varepsilon_n n! (2\pi i)^{-1} A_n(F)$$

derives from (5.36) and (5.37), from which

$$|\widetilde{A}_n(F)|2^{-n}(n!)^{-1} = \varepsilon_n(2\pi)^{-1}|A_n(F)|2^{-n}.$$

Therefore,

$$\widetilde{R}^{-1} = \limsup_{n \to \infty} \left(|A_n(F)| \, 2^{-n} \right)^{\frac{1}{n}} = (2R)^{-1},$$

i.e.

$$\widetilde{R} = 2R. \tag{5.38}$$

If we suppose that $R = \infty$, then $\tilde{R} = \infty$. Therefore, F should be an entire function, and since $F(\infty) = 0$, then $F = 0$. The obtained contradiction proves that the series (5.33) has a finite radius of convergence. $\qquad\square$

Lemma 5.7. *Let G be a simply connected domain in the right half-plane $\{z : z \in \mathbb{C}, \Re(z) > 0\}$, $\gamma \subset G$ be a rectifiable Jordan curve, $F \in H_\gamma$, F be not identically zero and the function $f(w)$ be defined by (5.32). Then a constant $R > 0$ exists, such that the function $f(w)$ is a holomorphic one in the half-plane $\Re(w) > -R$ $(R < \infty)$.*

Proof. Let $\inf_{z \in \gamma} \Re(z) = \rho$ and $\Re(w) > -\rho/2$. Then $\Re(z + 2w) > 0$ and therefore, the function Φ, given by (2.20), is holomorphic, from which the function $f(w)$, defined by (5.32), is holomorphic, too. Denote

$$R = \max\left\{\frac{\rho}{2} : f(w) \text{ is holomorphic, } \Re(w) > -\frac{\rho}{2}\right\}.$$

Then the function $f(w)$ is holomorphic for $\Re(w) > -R$. The assumption $R = \infty$ together with (5.38) imply that F is an entire function and therefore, $F = 0$. It shows that $R < \infty$ and $f(w)$ is holomorphic in a half-plane. $\qquad\square$

5.7. Theorems about Completeness for Neumann Polynomials Systems

Theorem 5.8. *Let $G \subset \mathbb{C}\backslash\{0\}$ be a simply connected region. Then the system $\{O_n(z)\}_{n=0}^\infty$ is complete in the space $H(G)$.*

Proof. Let $\gamma \subset G$ be a rectifiable Jordan curve, $F \in H_\gamma$ and

$$\int_\gamma F(z)O_n(z)dz = 0, \quad n = 0, 1, 2, \ldots. \tag{5.39}$$

By induction it can be proved that

$$\int_\gamma F(z)z^{-n}dz = 0, \quad n = 0, 1, 2, \ldots. \tag{5.40}$$

So, first let $n = 0$. Then, using (1.15), we obtain that

$$\int_\gamma F(z)O_0(z)dz = \int_\gamma F(z)z^{-1}dz = 0.$$

For $n = 1$, according to (1.13),

$$\int_\gamma F(z)O_1(z)dz = \int_\gamma F(z)z^{-2}dz = 0.$$

Suppose that the equality in (5.40) is fulfilled for $n = 0, 1, 2, \ldots, 2k$, i.e.

$$\int_\gamma F(z)z^{-n}dz = 0, \quad n = 0, 1, 2, \ldots, 2k. \tag{5.41}$$

Then, considering $\int_\gamma F(z)O_{2k}(z)dz$ and $\int_\gamma F(z)O_{2k+1}(z)dz$ and using (1.12), (1.13), (5.39) and (5.41), we have

$$0 = \int_\gamma F(z)O_{2k}(z)dz$$

$$= \frac{k + \frac{1}{2}}{2} \int_\gamma F(z) \sum_{m=0}^{k} \frac{(k+m-1)!}{(k-m)!} \left(\frac{z}{2}\right)^{-2m-1} dz$$

$$= \frac{(2k-1)!\,k}{2} \int_\gamma F(z) \left(\frac{z}{2}\right)^{-2k-1} dz.$$

Analogously,

$$0 = \int_\gamma F(z)O_{2k+1}(z)dz$$

$$= \frac{k + \frac{1}{2}}{2} \int_\gamma F(z) \sum_{m=0}^{k} \frac{(k+m)!}{(k-m)!} \left(\frac{z}{2}\right)^{-2m-2} dz$$

$$= \frac{(2k)!\left(k + \frac{1}{2}\right)}{2} \int_\gamma F(z) \left(\frac{z}{2}\right)^{-2k-2} dz.$$

Hence, $\int_\gamma F(z)z^{-2k-1}dz = 0$ and $\int_\gamma F(z)z^{-2k-2}dz = 0$, from which the validity of (5.40) holds.

At last, since the system $\{z^{-n}\}_{n=0}^{\infty}$ is complete in the space $H(G)$, then the equalities (5.40) imply the identity $F = 0$. Then, because of (5.39), the system $\{O_n(z)\}_{n=0}^{\infty}$ is complete in the space $H(G)$.

$\square$

Theorem 5.9. *Let* $\lim_{n \to \infty}(n/k_n) = \delta > 0$. *Then the system*

$$\{O_{k_n}(z)\}_{n=0}^{\infty} \tag{5.42}$$

is complete in the space $H(G)$ *for every simply connected region* G *in the right half-plane* $\Re(z) > 0$.

Proof. Let us suppose that $\lim_{n \to \infty}(n/k_n) = \delta > 0$, but nevertheless there exists a simply connected region G in the right half-plane, such that the family $\{O_{k_n}(z)\}_{n=0}^{\infty}$ is not complete in the space $H(G)$. This means that there exist a rectifiable Jordan curve $\gamma \subset G$ and a function $F \in H_\gamma$, F is not identically zero but

$$\int_\gamma F(z)O_{k_n}(z)dz = 0, \quad n = 0, 1, 2, \ldots.$$

Let us note that the function $f(w)$ is not identically zero (the opposite one, because of the completeness of the whole family of functions (Theorem 5.8), would mean that $F = 0$). Moreover, if Δ is the maximal density of the increasing sequence of indices of the nonzero coefficients (5.34) in the series (5.33), then $\delta > 0$ implies

$$\Delta \leq 1 - \delta < 1. \tag{5.43}$$

Since F is not identically zero, then, according to Theorem 5.8, not all of the complex numbers (5.34) are equal to zero. The power series (5.33) has a finite radius R of convergence ($0 < R < \infty$) and $w = -R$ is the unique singular point of $f(w)$, belonging to the circle $|w| = R$.

On the other hand, according to the Pólya Theorem 5.1, every closed arc of the circle $|w| = R$ with corresponding central angle $2\pi\Delta$ contains at least one singular point of $f(w)$. Following (5.43) we have $2\pi\Delta < 2\pi$, that leads us to a contradiction. $\square$

Corollary 5.3. *Let* $\lim_{n\to\infty}(n/k_n) = \delta > 0$. *Then the system*

$$\left\{ O_{k_n}\left(\frac{1}{z}\right) \right\}_{n=0}^{\infty} \tag{5.44}$$

is complete in the space $H(G)$ *for every simply connected region* G *in the right half-plane* $\Re(z) > 0$.

Proof. The completeness of the system (5.44) follows according to Theorem 5.9, after changing of z with z^{-1}. $\qquad\square$

Chapter 6

Multi-index Bessel Functions

6.1. Multi-index Generalizations of Bessel Functions of the First Kind

The Bessel functions and their various generalizations, originating from concrete problems in mechanics and astronomy, have proved themselves as some of the most frequently used special functions in mathematical analysis and its applications in physics, mechanics and engineering.

Let us first mention the so-called Bessel–Clifford functions $C_\nu(z)$, closely related to the Bessel functions $J_\nu(z)$,

$$C_\nu(z) = z^{-\frac{\nu}{2}} J_\nu(2\sqrt{z}) = \sum_{k=0}^{\infty} \frac{(-1)^k (z)^k}{k!\Gamma(\nu+k+1)}, \quad \nu \in \mathbb{C}, \qquad (6.1)$$

which are entire functions of z. Generalizations of the Bessel functions (more precisely, of the Bessel–Clifford functions) involving one more additional index μ have been introduced by Wright [Wright (1933)] and called Bessel–Wright functions or also misnamed in the literature as Bessel–Maitland functions (after Sir Edward Maitland Wright), namely:

$$J_\nu^\mu(z) = \sum_{k=0}^{\infty} \frac{(-z)^k}{k!\Gamma(\nu+\mu k+1)}, \quad \mu > -1, \qquad (6.2)$$

for details, see [Marichev (1978, p. 109); Kiryakova (1994, p. 336)], and so on. Initially, Wright defined (6.2) only for $\mu > 0$, and on

93

a later stage extended its definition to $\mu > -1$ (see for example [Kiryakova (1994, 2010a)]). More general are the three- and four-index generalizations of the Bessel function J_ν, namely generalized Bessel–Maitland (or Wright's) functions introduced by Pathak [1966] (for details, see also [Kiryakova (2010a and 2010b)]):

$$J_{\nu,\lambda}^{\mu}(z) = \left(\frac{z}{2}\right)^{\nu+2\lambda} \sum_{k=0}^{\infty} \frac{(-1)^k \left(\frac{z}{2}\right)^{2k}}{\Gamma(\lambda+k+1)\Gamma(\nu+k\mu+\lambda+1)},$$

$$z \in \mathbb{C}\backslash(-\infty,0], \ \mu > 0, \ \nu,\lambda \in \mathbb{C}, \qquad (6.3)$$

and the generalized Lommel–Wright functions, introduced by de Oteiza *et al.* (for details and results related to fractional calculus (FC), see [Kiryakova (2010b)] and also [Prieto *et al.* (2007)]),

$$J_{\nu,\lambda}^{\mu,m}(z) = \left(\frac{z}{2}\right)^{\nu+2\lambda} \sum_{k=0}^{\infty} \frac{(-1)^k \left(\frac{z}{2}\right)^{2k}}{(\Gamma(\lambda+k+1))^m \, \Gamma(\nu+k\mu+\lambda+1)}$$

$$z \in \mathbb{C}\backslash(-\infty,0], \ \mu > 0, \ m \in \mathbb{N}, \ \nu,\lambda \in \mathbb{C}. \qquad (6.4)$$

One more interesting generalization is the *hyper-Bessel function* $J_{\nu_1,\ldots,\nu_m}^{(m)}$, defined by the formula

$$J_{\nu_1,\ldots,\nu_m}^{(m)}(z) = \frac{\left(\frac{z}{m+1}\right)^{\nu_1+\cdots+\nu_m}}{\Gamma(\nu_1+1)\ldots\Gamma(\nu_m+1)} j_{\nu_1,\ldots,\nu_m}^{(m)}(z), \qquad (6.5)$$

where $z \in \mathbb{C}\backslash(-\infty,0]$, $\nu_i \in \mathbb{C}$, $\Re(\nu_i+1) > 0 \ (i = 1,\ldots,m)$, and

$$j_{\nu_1,\ldots,\nu_m}^{(m)}(z) = \sum_{k=0}^{\infty} \frac{(-1)^k \left(\frac{z}{m+1}\right)^{k(m+1)}}{(\nu_1+1)_k \ldots (\nu_m+1)_k} \frac{1}{k!}, \quad |z| < \infty. \qquad (6.6)$$

In view of (6.6), the hyper-Bessel functions can be written in the form:

$$J_{\nu_1,\ldots,\nu_m}^{(m)}(z) = \left(\frac{z}{m+1}\right)^{\sum_{i=1}^{m}\nu_i}$$

$$\times \sum_{k=0}^{\infty} \frac{(-1)^k \left(\frac{z}{m+1}\right)^{k(m+1)}}{\Gamma(k+\nu_1+1)\ldots\Gamma(k+\nu_m+1)} \frac{1}{k!}, \qquad (6.7)$$

being defined for $z \in \mathbb{C}\backslash(-\infty,0]$.

In 1953, this function was introduced by Delerue [1953] as a natural generalization of order m with vector indices $\nu = (\nu_1, \nu_2, \ldots, \nu_m)$ (or with multi-index $(\nu_1, \ldots, \nu_m)$) of the Bessel function of the first type J_ν. Later, this function was also studied by other authors, for example, by Marichev (1978), Dimovski (1966), Kljuchantzev (1975, 1987), Dimovski and Kiryakova (1986, 1987), Kiryakova (1994, 1997, 2014), Paneva-Konovska (2014a), and many others.

The hyper-Bessel functions of Delerue are closely related to the *hyper-Bessel differential operators* of arbitrary order $m > 1$, *introduced by Dimovski* [1966]. These are singular linear differential operators that appear very often in the problems of mathematical physics as a generalization of the second-order Bessel operator that can be represented in the alternative forms:

$$B = z^{\alpha_0} \frac{d}{dz} z^{\alpha_1} \cdots \frac{d}{dz} z^{\alpha_m} = z^{-\beta} \prod_{k=1}^{m} \left(z\frac{d}{dz} + \beta\gamma_k \right)$$

$$= z^{-\beta} \left(z^m \frac{d^m}{dz^m} + a_1 z^{m-1} \frac{d^{m-1}}{dz^{m-1}} + \cdots + a_{m-1} z\frac{d}{dz} + a_m \right), \quad (6.8)$$

$0 < z < \infty$, with sets of $(m + 1)$ parameters $\{\alpha_0, \alpha_1, \ldots, \alpha_m\}$, or $\{\beta > 0, \gamma_k$ real, $k = 1, \ldots, m\}$, or $\{\beta > 0, a_1, \ldots, a_m\}$. For details, see also [Dimovski and Kiryakova (1986, 1987); Kiryakova (1994, Chapter 3)]. Therefore, as shown in Theorem 3.4.3 and Corollary 3.4.4 in [Kiryakova (1994)], the fundamental system of solutions of the mth-order *hyper-Bessel differential equation $By(z) = \lambda y(z)$*, $\lambda \neq 0$, consists of the set of hyper-Bessel functions

$$J_{1+\gamma_1-\gamma_k,\ldots,*,\ldots,1+\gamma_m-\gamma_k}^{(m-1)} \left[(-\lambda)^{\frac{1}{m}} \left(\frac{m}{\beta} \right) z^{\frac{\beta}{m}} \right], \quad k = 1, \ldots, m,$$

under assumption of formal arrangement of the γ-parameters as $\gamma_1 < \gamma_2 < \cdots < \gamma_m < \gamma_1 + 1$ and where $*$ means to omit the kth term in the indices. And then, the solutions of hyper-Bessel ordinary differential equations (ODEs) $By(z) = \lambda y(z) + f(z)$ can be given explicitly in terms of hyper-Bessel functions, series in them, or series in integrals of them [Kiryakova (1994)].

Evidently, the hyper-Bessel functions (6.7) are natural generalizations of the *Bessel function* of the first kind (with $m + 1 = 2$,

$m = 1$), i.e.

$$J_\nu^{(1)}(z) = J_\nu(z) = \left(\frac{z}{2}\right)^\nu \sum_{k=0}^\infty \frac{(-1)^k \left(\frac{z}{2}\right)^{2k}}{\Gamma(k+\nu+1)} \frac{1}{k!}, \qquad (6.9)$$

as well as of the so-called *Bessel–Clifford functions of third-order* ($m + 1 = 3$): $C_{\nu,\mu}(z)$, depending on two ($m = 2$) indices and modifying the hyper-Bessel functions $J_{\nu,\mu}^{(2)}(z)$, namely

$$C_{\nu,\mu}(z) = z^{-\frac{\mu+\nu}{3}} J_{\nu,\mu}^{(2)}(3\sqrt[3]{z}) = \sum_{k=0}^\infty \frac{(-1)^k (z)^k}{\Gamma(k+\mu+1)\Gamma(k+\nu+1)} \frac{1}{k!}, \tag{6.10}$$

see details in [Kiryakova and Hernandez-Suarez (1995)]. Other interesting special cases of the hyper-Bessel functions of arbitrary order but with specific choice of indices are the so-called *trigonometric functions of order* m, including the *generalized* $\cos_m$- and the *generalized* $\sin_{m,k}$-*functions* ($k = 1, \ldots, m - 1$), studied for example by Kljuchantzev [1975, 1987], Dimovski and Kiryakova [1986, 1987], Kiryakova [1994, 1997], and many others. They appear as solutions of well-known classical case with particular hyper-Bessel operator $B = \left(\frac{d}{dz}\right)^m$. For example, the solution of the Cauchy problem

$$y^{(m)}(z) = -y(z), \quad y(0) = 1, \quad y'(0) = \cdots = y^{(m-1)}(0) = 0,$$

is given by

$$y(z) = \cos_m(z) = \sum_{k=0}^\infty \frac{(-1)^k z^{mk}}{(mk)!} = j_{\nu_1,\ldots,\nu_{m-1}}^{(m-1)}(z) \quad \text{with } \nu_k := \frac{k}{m} - 1.$$

At last, in particular (see e.g. [Marichev (1978, p. 110); Kiryakova (1994, pp. 352–353)]):

$$J_\nu^1(z) = C_\nu(z), \quad J_{\nu,0}^1(z) = J_\nu^{(1)}(z) = J_\nu(z),$$

$$J_{\nu,0}^{\mu}(z) = \left(\frac{z}{2}\right)^{\nu} J_{\nu}^{\mu}\left(\frac{z^2}{4}\right),$$

$$J_{\nu,\lambda}^{\mu,1}(z) = J_{\nu,\lambda}^{\mu}(z), \quad J_{\nu,\lambda}^{1}(z) = \frac{2^{2-2\lambda-\nu}}{\Gamma(\lambda)\Gamma(\lambda+\nu)} s_{2\lambda+\nu-1,\nu}(z),$$

$$\mathbf{H}_{\nu}(z) = \frac{2^{1-\nu}}{\sqrt{\pi}\Gamma\left(\nu+\frac{1}{2}\right)} s_{\nu,\nu}(z) = J_{\nu,\frac{1}{2}}^{1}(z),$$

$$(6.11)$$

where $s_{\alpha,\nu}(z)$ and $\mathbf{H}_{\nu}(z)$ denote respectively the Lommel and Struve functions [Erdélyi *et al.* (1953, Vol. 2, 7.5.5 (69), (84))].

Note that, generally speaking, this chapter is mainly based on the results obtained in [Paneva-Konovska (2007, 2008, 2010b, 2014a)].

6.2. Results on the Parameters of the Multi-index Generalizations of Bessel Functions

Now, consider the generalized Lommel–Wright functions (6.4) for

$$\mu > 0, \quad m \in \mathbb{N},$$

and the other indices of the kind $\nu = n - 2\lambda$, $n = 0, 1, 2, \ldots$, namely:

$$J_{n-2\lambda,\lambda}^{\mu,m}(z) = \left(\frac{z}{2}\right)^n \sum_{k=0}^{\infty} \frac{(-1)^k \left(\frac{z}{2}\right)^{2k}}{(\Gamma(\lambda+k+1))^m \Gamma(n-\lambda+k\mu+1)}, \quad z \in \mathbb{C}.$$

$$(6.12)$$

It is not hard to verify that the function $J_{n-2\lambda,\lambda}^{\mu,m}$ is entire (see e.g. [Kiryakova (2000)]).

Remark 6.1. In what follows, we will use the notations $\mathbb{R}^-$ (resp. $\mathbb{R}^+$) for the set of negative (resp. positive) real numbers, $\mathbb{Z}^-$ (resp. $\mathbb{N}$) for the set of negative (resp. positive) integers and $\mathbb{Z}_0^- = \mathbb{Z}^- \cup \{0\}$ (resp. $\mathbb{N}_0 = \mathbb{N} \cup \{0\}$).

Remark 6.2. Given a number λ, suppose that some coefficients in (6.12) are equal to zero, that is, there exist numbers $p \in \mathbb{N}_0$ and

$s \in \mathbb{N}$ such that the identity (6.12) can be written as

$$J_{n-2\lambda,\lambda}^{\mu,m}(z) = \left(\frac{z}{2}\right)^n \left(\frac{(-1)^p \left(\frac{z}{2}\right)^{2p}}{(\Gamma(\lambda+p+1))^m \Gamma(n-\lambda+p\mu+1)} \right.$$

$$\left. + \sum_{k=p+s}^{\infty} \frac{(-1)^k \left(\frac{z}{2}\right)^{2k}}{(\Gamma(\lambda+k+1))^m \Gamma(n-\lambda+k\mu+1)} \right). \quad (6.13)$$

Further, setting

$$a_k = \frac{1}{\Gamma(\lambda+k+1)}, \quad b_k = \frac{1}{\Gamma(n-\lambda+k\mu+1)}, \quad c_k = a_k{}^m b_k,$$

$$(6.14)$$

for all values of $k = 0, 1, 2, \ldots$, we consider two main cases, depending on whether the parameter λ is a positive integer or it is not. The results are given in both lemmas below. Beginning with $\lambda \notin \mathbb{N}$, the discussion separately goes in three cases as follows.

Lemma 6.1. *If $\lambda \in \mathbb{C}$, but λ is not a positive integer, then*

(i) $p = 0$ and $s = 1$ for $\lambda \notin (\mathbb{Z}^- \cup \mathbb{R}^+)$,
(ii) $p = -\lambda$ and $s = 1$ for $\lambda \in \mathbb{Z}^-$,
(iii) $p = 0$ and $s = 1$ or $s = 2$ for $\lambda \in \mathbb{R}^+$.

Proof. (i) Obviously, in this case, $\lambda \in \mathbb{C} \backslash (\mathbb{Z}^- \cup \mathbb{R}^+)$, that means $\lambda \leq 0$ but $\lambda \notin \mathbb{Z}^-$ or $\lambda \notin \mathbb{R}$. Letting $\lambda \leq 0$ and $\lambda \neq -1, -2, \ldots$, we conclude that both $\lambda + k + 1$ and $n - \lambda + k\mu + 1$ are neither negative integers nor zero and because of that $c_k \neq 0$, for all the values of k. Taking $\lambda \notin \mathbb{R}$, we deduce that $\lambda + k + 1 \notin \mathbb{R}$, as well as $n - \lambda + \mu k + 1 \notin \mathbb{R}$. That is why $c_k \neq 0$ as well. So, since all the coefficients are different from zero, then $p = 0$ and $s = 1$.

(ii) Now, let λ be a negative integer. Then $(n - \lambda + k\mu + 1)$ are positive numbers for all values of k and the integers $\lambda + k + 1 \notin \mathbb{Z}_0^-$ for $k \geq -\lambda$, id est, $c_k = 0$ for $0 \leq k < -\lambda$ and $c_k \neq 0$ for all $k \geq -\lambda$, that means $p = -\lambda$ and $s = 1$.

(iii) Finally, if λ is a positive noninteger real number, then a_k are positive numbers for each $k = 0, 1, 2, \ldots$ and $b_0 \neq 0$. However, it is possible that some of the rest b_k to be zero. Now, letting $k = 1$, then two possibilities exist: either $n - \lambda + \mu + 1 \notin \mathbb{Z}_0^-$, and then $b_1 \neq 0$, or $n - \lambda + \mu + 1 \in \mathbb{Z}_0^-$, and then $b_1 = 0$. The second case is possible only if $\mu \notin \mathbb{N}$, but then $n - \lambda + 2\mu + 1 \notin \mathbb{Z}_0^-$ and $b_2 \neq 0$. So, there are two possibilities: $c_0 \neq 0$ and $c_1 \neq 0$, or $c_0 \neq 0$ and $c_1 = 0$, but $c_2 \neq 0$. $\qquad\square$

Alternatively, for λ being a positive integer, the following statement can be made.

Lemma 6.2. *If λ is a positive integer, then*

(i) $p = 0$ *and* $s = 1$ *for* $n \geq \lambda$,
(ii) $p = 1$ *and* $s = 1$ *or* $s = 2$ *for* $0 \leq n < \lambda, \mu \notin \mathbb{N}$,
(iii) $p = \left[\frac{\lambda - n - 1}{\mu}\right] + 1$ *and* $s = 1$ *for* $0 \leq n < \lambda, \mu \in \mathbb{N}$.

Proof. First, let $n \geq \lambda$. Then $n - \lambda + 1 > 0, \lambda + k + 1 > 0$ and therefore, all coefficients $c_k \neq 0$. Now, let $0 \leq n < \lambda$. Then all a_k are positive but $n - \lambda + 1 \in \mathbb{Z}_0^-$ and because of that $b_0 = 0$. Further, if $\mu \notin \mathbb{N}$, then $b_1 \neq 0$ and, like in Lemma 6.1 case (iii), $b_2 \neq 0$ or $b_2 = 0$, but $b_3 \neq 0$. If $\mu \in \mathbb{N}$, then $n - \lambda + k\mu + 1 \in \mathbb{Z}$ and therefore, $b_k = 0$ for $k \leq \frac{\lambda - n - 1}{\mu}$, that is, $b_k > 0$ for $k > \frac{\lambda - n - 1}{\mu}$. So, $c_k = 0$ for $0 \leq k \leq \left[\frac{\lambda - n - 1}{\mu}\right]$, and $c_k \neq 0$ for $k \geq \left[\frac{\lambda - n - 1}{\mu}\right] + 1$. $\qquad\square$

Remark 6.3. Note that the relation (6.13), applied for $m = 1$, in view of the equalities (6.3), (6.11) and (6.14), leads to the representation of the function $J_{n-2\lambda,\lambda}^{\mu}(z)$, with the same p and s as the function $J_{n-2\lambda,\lambda}^{\mu,m}(z)$. For the case $m = 1$, $\lambda = 0$ yields the respective representation for the Bessel–Maitland functions with $p = 0$ and $s = 1$. We just mention that all the coefficients in (6.2) are different from zero, when $\nu \in \mathbb{N}_0$ and $\mu > 0$ (*cf.* with Lemma 6.1, case (i), applied for $\lambda = 0$).

Remark 6.3 along with the lemmas, proven above, show that the functions $J_n^{\mu}(z)$ and $J_{n-2\lambda,\lambda}^{\mu,m}(z)$ can be written in the forms:

$$J_n^{\mu}(z) = \frac{1}{\Gamma(n+1)}\left(1 + \vartheta_n^{\mu}(z)\right), \quad \mu > 0, \qquad (6.15)$$

with

$$\vartheta_n^\mu(z) = \sum_{k=1}^{\infty} \frac{\Gamma(n+1)(-z)^k}{k!\Gamma(n+\mu k+1)}, \qquad (6.16)$$

and respectively

$$J_{n-2\lambda,\lambda}^{\mu,m}(z) = \frac{(-1)^p \left(\frac{z}{2}\right)^{n+2p}}{(\Gamma(\lambda+p+1))^m \, \Gamma(n-\lambda+p\mu+1)} \left(1 + \vartheta_{n-2\lambda,\lambda}^{\mu,m}(z)\right), \qquad (6.17)$$

with

$$\vartheta_{n-2\lambda,\lambda}^{\mu,m}(z) = \sum_{k=p+s}^{\infty} \frac{(\Gamma(\lambda+p+1))^m \Gamma(n-\lambda+p\mu+1)}{(\Gamma(\lambda+k+1))^m \, \Gamma(n-\lambda+k\mu+1)} \left(-\frac{z^2}{4}\right)^{k-p}. \qquad (6.18)$$

The parameters p and s ($p \geq 0, s \geq 1$) are determined through the previous lemmas.

Now, let us come back to the hyper-Bessel functions. For further considerations relating to them, we construct a suitable enumerable family in order to obtain the most simple results.

To select such a family, let us fix an index $1 \leq i_0 \leq m$ and consider hyper-Bessel functions with integer indices $\nu_{i_0} = n$ ($n = 0, 1, 2, \ldots$). The use of n instead of ν_{i_0} is for simplifying the further exposition. Now, for the sake of brevity, set

$$\sum_{i=1}^{m}{}' \nu_i = \sum_{i=1, i\neq i_0}^{m} \nu_i, \quad \prod_{i=1}^{m}{}' \Gamma(k+\nu_i+1) = \prod_{i=1, i\neq i_0}^{m} \Gamma(k+\nu_i+1),$$

$$(\nu_{i_0}(n)) = (\nu_1, \ldots, \nu_{i_0-1}, n, \nu_{i_0+1}, \ldots, \nu_m) = (\nu_i)|_{\nu_{i_0}=n}. \qquad (6.19)$$

Then the functions of the discussed family have indices of the kind $(\nu_{i_0}(n))$ and, according to (6.7), they can be written in the form:

$$J_{(\nu_{i_0}(n))}^{(m)}(z) = A(z;m,n) \sum_{k=0}^{\infty} \frac{(-1)^k \left(\frac{z}{m+1}\right)^{k(m+1)}}{\Gamma(k+n+1) \prod_{i=1}^{m}{}' \Gamma(k+\nu_i+1)} \frac{1}{k!}, \qquad (6.20)$$

with

$$A(z; m, n) = \left(\frac{z}{m+1}\right)^{\sum\limits_{i=1}^{m}{}'\nu_i} \left(\frac{z}{m+1}\right)^n, \quad z \in \mathbb{C}\backslash(-\infty, 0].$$

(6.21)

Remark 6.4. Let us point out that the conditions $\Re(\nu_i + 1) > 0$ and $\Re(n + 1) > 0$ $(i = 1, \ldots, m)$, imposed on the given parameters, lead to the conclusion that the coefficients of (6.20) are all different from zero.

6.3. Inequalities Related to the Generalizations of the Bessel Functions

In this section, we intend to estimate the absolute value of the entire functions (6.16) and (6.18) in the complex plane $\mathbb{C}$ and its compact subsets, when $\mu > 0$. To this end, we need to know well the basic properties of the Γ-function.

Remark 6.5. Recall, for using in the further considerations, that

(i) Euler's Γ-function is positive in the interval $(0, \infty)$. Moreover, there exists a number $1 < \alpha_0 < 2$, such that the Euler's gamma function $\Gamma(\alpha)$ increases in (α_0, ∞), it decreases in $(0, \alpha_0)$ and in the set $\mathbb{R}^+$ of the positive real numbers, $\Gamma(\alpha)$ has its absolute minimum at the point α_0, i.e.

$$\min_{\alpha \in (0, \infty)} \Gamma(\alpha) = \Gamma(\alpha_0) > 0, \quad \alpha_0 \in (1, 2).$$

(ii) Stirling's formula says that

$$\Gamma(z + \alpha) \sim \sqrt{2\pi} z^{z + \alpha - \frac{1}{2}} \exp(-z), \quad |z| \to \infty,$$

with $|\arg(z + \alpha)| < \pi$.

(iii) Γ-functions quotient property holds, namely

$$\frac{\Gamma(z)}{\Gamma(z + \alpha)} = O\left(\frac{1}{z^\alpha}\right),$$

for $|\arg(z)| < \pi$ and $|\arg(z + \alpha)| < \pi$.

Beginning with $\vartheta_n^\mu(z)$, we give a result, referring to an upper estimate of its module, as follows.

Theorem 6.1. *Let $\vartheta_n^\mu(z)$ be the function, given by (6.16), then there exists a constant c_0, such that the following inequalities hold:*

$$|\vartheta_0^\mu(z)| \leq \frac{c_0}{\Gamma(1+\mu)} \left(\exp|z| - 1\right), \qquad (6.22)$$

respectively

$$|\vartheta_n^\mu(z)| \leq \frac{\Gamma(n+1)}{\Gamma(n+1+\mu)} \left(\exp|z| - 1\right) \quad (n \in \mathbb{N}), \qquad (6.23)$$

in the whole complex plane. Moreover, if $K \subset \mathbb{C}$ is a nonempty compact set, then a constant $0 < C = C(K) < \infty$ exists, such that

$$|\vartheta_n^\mu(z)| \leq C\Gamma(n+1)/\Gamma(n+1+\mu). \qquad (6.24)$$

for each $n \in \mathbb{N}_0$ and each $z \in K$.

Proof. Letting $z \in \mathbb{C}$, we denote for convenience

$$w_k(n,\mu) = \frac{\Gamma(n+\mu+1)}{\Gamma(n+\mu k+1)}, \quad u_k(z;n,\mu) = \frac{w_k(n,\mu)}{k!}(-z)^k. \quad (6.25)$$

Then, the function $\vartheta_n^\mu(z)$, given by the relation (6.16), can be written in the form:

$$\vartheta_n^\mu(z) = \frac{\Gamma(n+1)}{\Gamma(n+\mu+1)} \sum_{k=1}^{\infty} u_k(z;n,\mu). \qquad (6.26)$$

In order to estimate the module of $\vartheta_n^\mu(z)$, we need to find an upper estimation of $|w_k(n,\mu)|$.

First, let $n = 0$ and $1 < \mu+1 < \alpha_0$. Then, in view of Remark 6.5 case (i), both inequalities $\Gamma(\mu+1) < \Gamma(1) = 1$ and $\Gamma(\alpha_0) \leq \Gamma(\mu k+1)$ hold, from which $w_k(0,\mu) < 1/\Gamma(\alpha_0)$. If $\alpha_0 \leq \mu+1$, then the inequality $w_k(0,\mu) \leq 1$ holds. So, we obtain the estimate

$$|u_k(z;0,\mu)| \leq c_0 \frac{|z|^k}{k!}, \quad \text{with } c_0 = \begin{cases} \dfrac{1}{\Gamma(\alpha_0)} & \text{if } \mu+1 < \alpha_0, \\ 1 & \text{if } \mu+1 \geq \alpha_0, \end{cases}$$

for the absolute value of $u_k(z; n, \mu)$ and therefore, in view of (6.26), the estimation (6.22), concerning the module of $\vartheta_0^\mu(z)$, is obtained.

Further, let $n \in \mathbb{N}$. Then the inequalities $w_k(n, \mu) \le 1$ hold for all values of $k = 1, 2, \ldots$, and therefore, the validity of the inequality (6.23) is verified in the whole complex plane.

Now, the estimate (6.24) follows immediately in accordance with (6.22) and (6.23). □

Further, we are going to estimate the entire functions $\vartheta_{n-2\lambda,\lambda}^{\mu,m}(z)$. To this end, we firstly transform the expression in the equality (6.18) that leads to the identity

$$\vartheta_{n-2\lambda,\lambda}^{\mu,m}(z) = \widetilde{\Gamma} \sum_{k=p+s}^{\infty} \frac{(-1)^{k-p}\gamma_k}{(\Gamma(\lambda+k+1))^m} \left(\frac{z}{2}\right)^{2(k-p)}, \qquad (6.27)$$

with

$$\widetilde{\Gamma} = \frac{\Gamma^m(\lambda+p+1)\Gamma(n-\lambda+p\mu+1)}{\Gamma(n-\lambda+(p+s)\mu+1)},$$

$$\gamma_k = \frac{\Gamma(n-\lambda+(p+s)\mu+1)}{\Gamma(n-\lambda+k\mu+1)}. \qquad (6.28)$$

Subsequently that we give the results, separated in some different cases, depending on the parameter λ, as follows.

Lemma 6.3. *Let* $\lambda = 0$. *Then*

$$\left|\vartheta_{n-2\lambda,\lambda}^{\mu,m}(z)\right| \le \frac{\Gamma(n+1)}{\Gamma(n+\mu+1)} \left(\exp\left(\left|\frac{z^2}{4}\right|\right) - 1\right), \quad n \in \mathbb{N},$$

$$\left|\vartheta_{n-2\lambda,\lambda}^{\mu,m}(z)\right| \le \frac{1}{\Gamma(\alpha_0)} \left(\exp\left(\left|\frac{z^2}{4}\right|\right) - 1\right), \quad n = 0.$$

Proof. Because of Lemma 6.1 case (i), we have $p = 0, s = 1, \gamma_k = \frac{\Gamma(n+\mu+1)}{\Gamma(n+k\mu+1)}$ and

$$\vartheta_{n-2\lambda,\lambda}^{\mu,m}(z) = \frac{\Gamma(n+1)}{\Gamma(n+\mu+1)} \sum_{k=1}^{\infty} \frac{(-1)^k\gamma_k}{(\Gamma(k+1))^m} \left(\frac{z}{2}\right)^{2k},$$

from which

$$\left|\vartheta^{\mu,m}_{n-2\lambda,\lambda}(z)\right| \leq \frac{\Gamma(n+1)}{\Gamma(n+\mu+1)} \sum_{k=1}^{\infty} \frac{\left(\left|\frac{z}{2}\right|^2\right)^k}{(\Gamma(k+1))^m} \gamma_k$$

$$\leq \frac{\Gamma(n+1)}{\Gamma(n+\mu+1)} \sum_{k=1}^{\infty} \frac{\left(\left|\frac{z}{2}\right|^2\right)^k}{\Gamma(k+1)} \gamma_k.$$

Taking into consideration Remark 6.5 case (i), we can write $0 < \gamma_k \leq 1$ for $n \in \mathbb{N}$ and $0 < \gamma_k \leq \frac{\Gamma(\mu+1)}{\Gamma(\alpha_0)}$ for $n = 0$ that proves the lemma. $\qquad\square$

Lemma 6.4. *Let* $\lambda \in \mathbb{Z}^-$, *then*

$$\left|\vartheta^{\mu,m}_{n-2\lambda,\lambda}(z)\right| \leq \Gamma(1+\mu-\lambda\mu)\frac{\Gamma(n-\lambda-\lambda\mu+1)}{\Gamma(n-\lambda+(1-\lambda)\mu+1)}\xi\left(|z|^2/4; \lambda, \mu\right),$$

with

$$\xi\left(\frac{|z|^2}{4}; \lambda, \mu\right) = J^{\mu}_{-\lambda\mu}\left(-\left|\frac{z^2}{4}\right|\right) - \frac{1}{\Gamma(1-\lambda\mu)}.$$

Proof. According to Lemma 6.1 case (ii), $p = -\lambda$ and $s = 1$ and therefore, (6.27) takes the form:

$$\vartheta^{\mu,m}_{n-2\lambda,\lambda}(z) = \frac{\Gamma(n-\lambda-\lambda\mu+1)}{\Gamma(n-\lambda+(1-\lambda)\mu+1)}$$

$$\times \sum_{k=p+1}^{\infty} \frac{(-1)^{k-p}\gamma_k}{\Gamma^m(\lambda+k+1)} \left(\frac{z}{2}\right)^{2(k-p)},$$

with

$$\gamma_k = \frac{\Gamma(n-\lambda+(p+1)\mu+1)}{\Gamma(n-\lambda+k\mu+1)}.$$

Since $p+1 \leq k$ and

$$\gamma_k = \frac{(n-\lambda+(p+1)\mu)}{(n-\lambda+k\mu)} \times \cdots \times \frac{(1+(p+1)\mu)}{(1+k\mu)} \times \frac{\Gamma(1+(p+1)\mu)}{\Gamma(1+k\mu)},$$

then

$$0 < \gamma_k \leq \frac{\Gamma(1 + (1-\lambda)\mu)}{\Gamma(1 + k\mu)}.$$

Using denotation

$$w_k(z) = \frac{(-1)^{k-p}\gamma_k}{(\Gamma(\lambda + k + 1))^m} \left(\frac{z}{2}\right)^{2(k-p)},$$

we obtain consecutively

$$\sum_{k=p+1}^{\infty} w_k(z) = \sum_{k=1}^{\infty} w_{k+p}(z) = \sum_{k=1}^{\infty} \frac{(-1)^k \gamma_{k+p}}{(\Gamma(k+1))^m} \left(\frac{z}{2}\right)^{2k},$$

$$\left|\sum_{k=p+1}^{\infty} w_k(z)\right| \leq \sum_{k=1}^{\infty} \frac{\gamma_{k+p}\left(\left|\frac{z}{2}\right|^2\right)^k}{(\Gamma(k+1))^m} \leq \sum_{k=1}^{\infty} \frac{\Gamma(1 + \mu - \lambda\mu)\left(\left|\frac{z}{2}\right|^2\right)^k}{(\Gamma(k+1))^m \Gamma(1 + k\mu - \lambda\mu)}$$

$$= \Gamma(1 + \mu - \lambda\mu)$$

$$\times \left(\sum_{k=0}^{\infty} \frac{\left(\left|\frac{z}{2}\right|^2\right)^k}{(\Gamma(k+1))^m \Gamma(1 + k\mu - \lambda\mu)} - \frac{1}{\Gamma(1 - \lambda\mu)}\right)$$

$$\leq \Gamma(1 + \mu - \lambda\mu)$$

$$\times \left(\sum_{k=0}^{\infty} \frac{\left(\left|\frac{z}{2}\right|^2\right)^k}{\Gamma(k+1)\Gamma(1 + k\mu - \lambda\mu)} - \frac{1}{\Gamma(1 - \lambda\mu)}\right),$$

from which the conclusion of the theorem immediately follows. □

Lemma 6.5. *Let $\lambda \in \mathbb{R}^- \backslash \mathbb{Z}$, then there exists an entire function τ such that*

$$\left|\vartheta_{n-2\lambda,\lambda}^{\mu,m}(z)\right| \leq |\Gamma(\lambda + 1)|^m \frac{\Gamma(n - \lambda + 1)}{\Gamma(n - \lambda + \mu + 1)} \tau\left(\frac{|z|^2}{4}; \lambda\right).$$

Proof. Now, as a result of Lemma 6.1 case (i), we have $p = 0$, $s = 1$, $\gamma_k = \frac{\Gamma(n-\lambda+\mu+1)}{\Gamma(n-\lambda+k\mu+1)}$ and

$$\vartheta_{n-2\lambda,\lambda}^{\mu,m}(z) = \frac{\Gamma^m(\lambda+1)\Gamma(n-\lambda+1)}{\Gamma(n-\lambda+\mu+1)} \sum_{k=1}^{\infty} \frac{(-1)^k \gamma_k}{(\Gamma(\lambda+k+1))^m} \left(\frac{z}{2}\right)^{2k},$$

(6.29)

respectively

$$\left|\vartheta_{n-2\lambda,\lambda}^{\mu,m}(z)\right| \le |\Gamma(\lambda+1)| \frac{\Gamma(n-\lambda+1)}{\Gamma(n-\lambda+\mu+1)} \sum_{k=1}^{\infty} \frac{\left(\left|\frac{z}{2}\right|^2\right)^k}{|\Gamma(\lambda+k+1)|^m} \gamma_k.$$

Further, the increase of the gamma function leads to the inequality $0 < \gamma_k \le 1$ for both $n = 0$, but $1 - \lambda + \mu \ge \alpha_0$, and $n \in \mathbb{N}$, whereas $0 < \gamma_k \le \frac{1}{\Gamma(\alpha_0)}$ for $n = 0$ and $1 < 1 - \lambda + \mu < \alpha_0$. Finally, getting $C = \max\left(1, \frac{1}{\Gamma(\alpha_0)}\right) = \frac{1}{\Gamma(\alpha_0)}$ and $\tau(z;\lambda) = C \sum_{k=1}^{\infty} \frac{z^k}{|\Gamma(\lambda+k+1)|^m}$, we complete the proof of the lemma. $\square$

Lemma 6.6. *Let $\lambda \notin \mathbb{R}$, that is $\lambda = \lambda_1 + i\lambda_2$ ($\lambda_1, \lambda_2 \in \mathbb{R}$, $\lambda_2 \ne 0$), and let $\widetilde{\lambda}_1 = \max(0, \lambda_1)$. Then there exist entire functions φ and φ_1, such that*

$$\left|\vartheta_{n-2\lambda,\lambda}^{\mu,m}(z)\right| \le |\Gamma(\lambda+1)|^m \frac{|\Gamma(n-\lambda+1)|}{|\Gamma(n-\lambda+\mu+1)|} \varphi\left(\frac{|z|^2}{4}; \lambda, \mu\right),$$

for $n \ge \widetilde{\lambda}_1$, and

$$\left|\vartheta_{n-2\lambda,\lambda}^{\mu,m}(z)\right| \le |\Gamma(\lambda+1)|^m \frac{|\Gamma(n-\lambda+1)|}{|\Gamma(n-\lambda+\mu+1)|} \varphi_1\left(\frac{|z|^2}{4}; \lambda, \mu\right),$$

for $0 \le n < \widetilde{\lambda}_1$.

Proof. Analogously to Lemma 6.5, $p = 0$ and $s = 1$, from which the identity (6.29) expresses $\vartheta_{n-2\lambda,\lambda}^{\mu,m}(z)$, like in Lemma 6.5 and with the same γ_k.

First, we consider the case $n \geq \widetilde{\lambda}_1$. Following the idea of the proof of Lemma 6.4, we obtain the estimate

$$|\gamma_k| \leq \frac{|\Gamma([\widetilde{\lambda}_1] + 1 - \lambda_1 + \mu - i\lambda_2)|}{|\Gamma([\widetilde{\lambda}_1] + 1 - \lambda_1 + k\mu - i\lambda_2)|} = \frac{|\Gamma([\widetilde{\lambda}_1] + 1 - \lambda + \mu)|}{|\Gamma([\widetilde{\lambda}_1] + 1 - \lambda + k\mu)|},$$

where $[\widetilde{\lambda}_1]$ denotes the integer part of $\widetilde{\lambda}_1$.

Now, let $0 \leq n < \widetilde{\lambda}_1$. Then because of the convergence of the sequence $\{\gamma_k\}_{k=1}^{\infty}$, it is bounded and therefore, a constant $\widetilde{C}$ exists such that $|\gamma_k| \leq \widetilde{C}$ for all values of k.

Eventually, the proof ends taking

$$\varphi(z; \lambda, \mu) = \sum_{k=1}^{\infty} \frac{|\Gamma([\widetilde{\lambda}_1] + 1 - \lambda + \mu)|}{|(\Gamma(\lambda + k + 1))^m \Gamma([\widetilde{\lambda}_1] + 1 - \lambda + k\mu)|} z^k,$$

and

$$\varphi_1(z; \lambda, \mu) = \widetilde{C} \sum_{k=1}^{\infty} \frac{z^k}{|\Gamma(\lambda + k + 1)|^m}. \qquad \square$$

To formulate the rest two lemmas, we need the Mittag-Leffler function $E_{\alpha,\beta}(z)$. About the definition and basic properties of this entire function, see more in Chapter 7.

Lemma 6.7. *Let* $\lambda \in \mathbb{R}^+ \backslash \mathbb{N}$, *then*

$$\left| \vartheta_{n-2\lambda,\lambda}^{\mu,m}(z) \right| \leq (\Gamma(\lambda + 1))^m \Gamma([\lambda] - \lambda + s\mu + 1) \frac{\Gamma(n - \lambda + 1)}{\Gamma(n - \lambda + s\mu + 1)}$$

$$\times \left(\left(\frac{i|z|}{2} \right)^{-[\lambda]} J_{[\lambda]-2\lambda,\lambda}^{\mu,m}(i|z|) \right.$$

$$\left. - \sum_{k=0}^{s-1} \frac{\left| \frac{z}{2} \right|^{2k}}{\Gamma^m(\lambda + k + 1)\Gamma([\lambda] - \lambda + k\mu + 1)} \right),$$

for $n > \lambda$, *and there exist a constant* C *and an entire function* η *such that*

$$\left| \vartheta_{n-2\lambda,\lambda}^{\mu}(z) \right| \leq \frac{C\, \Gamma^m(\lambda + 1)\, |\Gamma(n - \lambda + 1)|}{|\Gamma(n - \lambda + s\mu + 1)|} \eta\left(\frac{|z|^2}{4}; \lambda \right),$$

for $0 \leq n < \lambda$. Moreover,

$$\eta\left(\frac{|z|^2}{4};\lambda\right) = E_{(1,\lambda+1)}\left(\frac{|z|^2}{4}\right) - \sum_{k=0}^{s-1}\frac{\left|\frac{z}{2}\right|^{2k}}{\Gamma(\lambda+k+1)},$$

when $m = 1$, and

$$\eta(|z|^2/4;\lambda) = \left(-\frac{|z|^2}{4}\right)^{-\lambda} J_{0,\lambda}^{1,m-1}(i|z|) - \sum_{k=0}^{s-1}\frac{\left|\frac{z}{2}\right|^{2k}}{\Gamma^m(\lambda+k+1)},$$

for $m > 1$.

Proof. According to Lemma 6.1 case (iii), we have $p = 0, s = 1$ or $s = 2$, $\gamma_k = \frac{\Gamma(n-\lambda+s\mu+1)}{\Gamma(n-\lambda+k\mu+1)}$ and

$$\vartheta_{n-2\lambda,\lambda}^{\mu,m}(z) = \frac{\Gamma^m(\lambda+1)\Gamma(n-\lambda+1)}{\Gamma(n-\lambda+s\mu+1)}\sum_{k=s}^{\infty}\frac{(-1)^k\gamma_k}{(\Gamma(\lambda+k+1))^m}\left(\frac{z}{2}\right)^{2k},$$

respectively

$$\left|\vartheta_{n-2\lambda,\lambda}^{\mu,m}(z)\right| \leq \Gamma^m(\lambda+1)\left|\frac{\Gamma(n-\lambda+1)}{\Gamma(n-\lambda+s\mu+1)}\right|\sum_{k=s}^{\infty}\frac{|\gamma_k|\left(\left|\frac{z}{2}\right|^2\right)^k}{(\Gamma(\lambda+k+1))^m}.$$

For $n > \lambda$, in the same way like in the proof of Lemma 6.6, we get to the inequality

$$0 < \gamma_k \leq \frac{\Gamma([\lambda]+1-\lambda+s\mu)}{\Gamma([\lambda]+1-\lambda+k\mu)}$$

for all $k \geq s$, that immediately ends the proof.

The case $n < \lambda$ goes again in the same way like in Lemma 6.6.

$\square$

Lemma 6.8. *Let $\lambda \in \mathbb{N}$, then*

$$\left|\vartheta_{n-2\lambda,\lambda}^{\mu,m}(z)\right| \leq \Gamma^m(\lambda+1)\Gamma(\mu+1)\frac{\Gamma(n-\lambda+1)}{\Gamma(n-\lambda+\mu+1)}\zeta\left(\frac{|z|^2}{4};\lambda,\mu\right),$$

with

$$\zeta\left(\frac{|z|^2}{4};\lambda,\mu\right) = \left(\frac{i|z|}{2}\right)^{-\lambda} J^{\mu}_{-\lambda,\lambda}(i|z|) - \frac{1}{\Gamma(\lambda+1)}$$

for all $n \geq \lambda$, and there exists a constant C such that for $0 \leq n < \lambda$

$$|\vartheta^{\mu,m}_{n-2\lambda,\lambda}(z)| \leq C\,\frac{\Gamma^m(\lambda+p+1)\,|\Gamma(n-\lambda+p\mu+1)|}{|\Gamma(n-\lambda+(p+s)\mu+1)|}\,\zeta_1\left(\frac{|z|^2}{4};\lambda,\mu\right)$$

holds, with

$$\zeta_1\left(\frac{|z|^2}{4};\lambda,\mu\right) = E_{(1,\lambda+1)}\left(\frac{|z|^2}{4}\right) - \sum_{k=0}^{p+s-1}\frac{\left|\frac{z}{2}\right|^{2k}}{\Gamma(\lambda+k+1)}$$

and the corresponding p and s.

Proof. Beginning with the case $n \geq \lambda$, we have, according to Lemma 6.2, that $p = 0$, $s = 1$ and the identity (6.29), like in the proof of Lemma 6.5, expresses $\vartheta^{\mu,m}_{n-2\lambda,\lambda}(z)$, with the same γ_k. Then, like in the proof of Lemma 6.4, we get to the inequalities $0 < \gamma_k \leq \frac{\Gamma(\mu+1)}{\Gamma(k\mu+1)}$ for all $k \in \mathbb{N}$ and therefore,

$$|\vartheta^{\mu,m}_{n-2\lambda,\lambda}(z)| \leq \frac{\Gamma^m(\lambda+1)\,\Gamma(n-\lambda+1)}{\Gamma(n-\lambda+\mu+1)}\sum_{k=1}^{\infty}\frac{\left(\left|\frac{z}{2}\right|^2\right)^k}{(\Gamma(\lambda+k+1))^m}\gamma_k.$$

Further, taking into account that $\Gamma^m(\lambda+k+1) \geq \Gamma(\lambda+k+1)$ and also

$$\zeta\left(\frac{|z|^2}{4};\lambda,\mu\right) = \sum_{k=0}^{\infty}\frac{\left(\left|\frac{z}{2}\right|^2\right)^k}{\Gamma(k\mu+1)\Gamma(\lambda+k+1)} - \frac{1}{\Gamma(\lambda+1)},$$

the desired estimation is verified, for all $n \geq \lambda$.

The proof of the case $n < \lambda$ proceeds in the same way, additionally using the convergence of the sequence $\{\gamma_k\}$ and its boundedness as well. $\qquad\square$

So, the complex plane can been separated into six subsets, non-intersecting one another, namely

$$\Delta_1 = \{0\}, \quad \Delta_2 = \mathbb{Z}^-, \quad \Delta_3 = \mathbb{R}^-\backslash\mathbb{Z},$$

$$\Delta_4 = \mathbb{C}\backslash\mathbb{R}, \quad \Delta_5 = \mathbb{R}^+\backslash\mathbb{N}, \quad \Delta_6 = \mathbb{N}$$

and $\mathbb{C} = \bigcup_{k=1}^6 \Delta_k$. We summarize the above-given lemmas formulating the following theorem.

Theorem 6.2. *Let $\vartheta_{n-2\lambda,\lambda}^{\mu,m}(z)$ be the function, given by (6.18), then there exist entire functions ψ_k $(1 \le k \le 6)$, such that the following inequalities hold:*

$$\left|\vartheta_{n-2\lambda,\lambda}^{\mu,m}(z)\right| \le \frac{|\Gamma(n - \lambda + p\mu + 1)|}{|\Gamma(n - \lambda + (p+s)\mu + 1)|}\psi_k\left(\frac{|z|^2}{4}; \lambda, \mu\right), \quad (6.30)$$

in the whole complex plane, provided $\lambda \in \Delta_k$.

Moreover, if $K \subset \mathbb{C}$ is a nonempty compact set, then a constant $0 < C = C(K) < \infty$ exists, such that

$$\left|\vartheta_{n-2\lambda,\lambda}^{\mu,m}(z)\right| \le C\frac{|\Gamma(n - \lambda + p\mu + 1)|}{|\Gamma(n - \lambda + (p+s)\mu + 1)|} \quad (6.31)$$

with the corresponding p and s, obtained in Lemmas 6.1 and 6.2, for each $n \in \mathbb{N}_0$ and each $z \in K$.

Proof. The establishment of the existence of $\psi_k(z; \lambda, \mu)$ and verity of (6.30) goes taking the function $\psi_k(|z|^2/4; \lambda, \mu)$ like in the preceding lemmas in this section, up to a constant multiplier. In accordance with the holomorphicity of $\psi_k(z; \lambda, \mu)$ in the whole complex plane $\mathbb{C}$, there exists a constant $C_k(K)$ such that $\left|\psi_k(|z|^2/4; \lambda, \mu)\right| \le C_k$ for each $z \in K$. Further, the validity of (6.31) follows immediately, setting $C = \max_{1\le k\le 6} C_k$. $\qquad\qquad\square$

Finally, we give a corresponding upper estimate for the hyper-Bessel functions (6.20).

Let $z \in \mathbb{C}\backslash(-\infty, 0]$, $\nu_i \in \mathbb{C}$, $\Re(\nu_i + 1) > 0$ $(i = 1, \dots, m)$ and $J_{(\nu_{i_0}(n))}$ $(n = 0, 1, \dots)$ be the functions (6.20). The following result refers to them.

Theorem 6.3. *There exists an entire function ϑ_n, such that the hyper-Bessel function $J^{(m)}_{(\nu_{i_0}(n))}$ can be written in the form:*

$$J^{(m)}_{(\nu_{i_0}(n))}(z) = \frac{A(z;m,n)}{\Gamma(n+1)\prod_{i=1}^{m}{}'\Gamma(\nu_i+1)}(1+\vartheta_n(z)), \qquad (6.32)$$

with $A(z;m,n)$ defined by (6.21) and

$$|\vartheta_n(z)| \leq \frac{1}{(n+1)\prod_{i=1}^{m}{}'|\nu_i+1|}\left(\frac{|z|}{m+1}\right)^{m+1}\exp\left(\left(\frac{|z|}{m+1}\right)^{m+1}\right),$$
$$(6.33)$$

for all the values of the variable z in the complex plane $\mathbb{C}$. Moreover, if K is a nonempty compact subset of $\mathbb{C}$, then there exists a constant $C = C(K)$, such that

$$|\vartheta_n(z)| \leq \frac{C}{n+1}, \qquad (6.34)$$

for each $n \in \mathbb{N}_0$ and each $z \in K$.

Remark 6.6. Essentially, the proof is exposed in Sec. 8.3.

6.4. Asymptotic Formulae with Respect to the Index n

In this section, we propose some asymptotic formulae with respect to the index for the generalized Bessel functions. Our results are natural generalizations of the known asymptotic formula (1.19) for the Bessel functions J_n with a nonnegative integer index n when $n \to \infty$.

Further, we consider the functions $J_n^\mu(z)$ and $J^{\mu,m}_{n-2\lambda,\lambda}(z)$ with integer indices $n = 0, 1, \ldots$, respectively defined by (6.2) and (6.12), for $\mu > 0$. Note that both functions $\vartheta_n^\mu(z)$ and $\vartheta^{\mu,m}_{n-2\lambda,\lambda}(z)$ as well as $J_n^\mu(z)$ and $J^{\mu,m}_{n-2\lambda,\lambda}(z)$ are holomorphic functions of z in the whole complex plane $\mathbb{C}$, i.e. they are all entire functions. In this section, we prove some asymptotic formulae, concerning the above-discussed Bessel type functions for 'large' values of the index n, beginning with the first of them.

Theorem 6.4. *Let $\mu > 0$. Then the Bessel–Maitland (Wright's) functions $J_n^\mu(z)$ have the asymptotic formula (6.15) with ϑ_n^μ defined by (6.16). Moreover, for all $z \in \mathbb{C}$,*

$$\vartheta_n^\mu(z) \to 0 \quad as \quad n \to \infty, \tag{6.35}$$

and the convergence is uniform on the compact subsets of the complex plane.

Proof. Recall that, according to Remark 6.5 case (iii), Stirling's formula gives the following property of the Γ-functions quotient, namely

$$\frac{\Gamma(n+1)}{\Gamma(n+1+\mu)} = O\left(\frac{1}{n^\mu}\right), \quad \text{for } n \in \mathbb{N}.$$

Then, since $n^{-\mu} \to 0$ and in view of (6.23), the validity of (6.35) follows immediately. The uniform convergence of $\{\vartheta_n^\mu(z)\}$ on the compact subsets of $\mathbb{C}$ follows from the estimate (6.24).   □

In an analogical way can be formulated the corresponding theorem referring to the generalized Lommel–Wright functions, given below.

Theorem 6.5. *Let $\mu > 0$. Then the Lommel–Wright functions $J_{n-2\lambda,\lambda}^{\mu,m}(z)$ have the asymptotic formula (6.17) with $\vartheta_{n-2\lambda,\lambda}^{\mu,m}$ defined by (6.18). Moreover, for all $z \in \mathbb{C}$,*

$$\vartheta_{n-2\lambda,\lambda}^{\mu,m}(z) \to 0 \quad as \quad n \to \infty, \tag{6.36}$$

and the convergence is uniform on the compact subsets of the complex plane.

Proof. In the same way as in the previous theorem, according to Remark 6.5 case (iii), the property of the Γ-functions quotient yields

$$\frac{\Gamma(n - \lambda + p\mu + 1)}{\Gamma(n - \lambda + (p+s)\mu + 1)} = O\left(\frac{1}{n^{s\mu}}\right), \quad \text{for } n \in \mathbb{N}.$$

Then, since the corresponding $s \geq 1$ (s is defined by Lemmas 6.1 and 6.2), the relation $n^{-s\mu} \to 0$ holds, and in view of (6.30), the correctness of (6.36) follows immediately. The uniform convergence of

$\left\{\vartheta_{n-2\lambda,\lambda}^{\mu,m}(z)\right\}$ on the compact subsets of $\mathbb{C}$ follows from the estimate (6.31). $\qquad\square$

Remark 6.7. Note that all the results, obtained in this chapter, concerning the generalized Lommel–Wright functions produce the analogical ones for the generalized Bessel–Maitland functions, taking $m = 1$.

Specifically, the following corollary can be made.

Corollary 6.1. *Let* $\mu > 0$. *Then the generalized Bessel–Maitland (Wright's) functions* (6.3) *have the following asymptotic formula*

$$J_{n-2\lambda,\lambda}^{\mu}(z)$$

$$= \frac{(-1)^p \left(\frac{z}{2}\right)^{n+2p}}{\Gamma(\lambda+p+1)\Gamma(n-\lambda+p\mu+1)} \left(1 + \vartheta_{n-2\lambda,\lambda}^{\mu}(z)\right), \quad z \in \mathbb{C},$$

$$\vartheta_{n-2\lambda,\lambda}^{\mu}(z) \to 0 \quad as \quad n \to \infty \quad (n \in \mathbb{N}). \tag{6.37}$$

The functions $\vartheta_{n-2\lambda,\lambda}^{\mu}(z)$ *are holomorphic functions of* z *in* $\mathbb{C}$. *The convergence is uniform on the compact subsets of the complex plane* $\mathbb{C}$.

Theorem 6.6. *The hyper-Bessel function* $J_{\left(\nu_{i_0}(n)\right)}^{(m)}$ *has the asymptotic formula, given by* (6.32), *where* $\vartheta_n(z) \to 0$ *when* $n \to \infty$. *Moreover, on the compact subsets of* $\mathbb{C}$ *the convergence of* $\vartheta_n(z)$ *is uniform.*

Proof. Let ϑ_n be the function in (6.32). Now, bearing in mind that the inequalities (6.33) and (6.34) hold, the confirmation of the result follows automatically. $\qquad\square$

Remark 6.8. According to Theorems 6.4–6.6 and Corollary 6.1, it follows that there exists a natural number M such that the functions $J_n^{\mu}(z)$, $J_{n-2\lambda,\lambda}^{\mu,m}(z)$, $J_{n-2\lambda,\lambda}^{\mu}(z)$ and $J_{\left(\nu_{i_0}(n)\right)}^{(m)}(z)$ have no zeros for $n > M$, possibly except for the zero.

6.5. Special Cases

Some interesting cases of the above-discussed Bessel type functions are given below. As written in (6.11), for $m = 1$ the special function (6.4) turns into the generalization (6.3) of the Bessel function $J_\nu(z)$, introduced by Pathak [1966] (for details also see [Kiryakova (2010a, 2010b)]), i.e.

$$J^\mu_{\nu,\lambda}(z) = J^{\mu,1}_{\nu,\lambda}(z). \tag{6.38}$$

For particular choices of the other parameters λ and μ, we obtain results for more special cases as follows.

(i) Let $\lambda = 0$, then the special function (6.3) gives the generalization of the Bessel–Clifford function $C_\nu(z) = z^{-\nu/2} J_\nu(2\sqrt{z})$, namely

$$J^\mu_\nu \left(\frac{z^2}{4} \right) = \left(\frac{z}{2} \right)^{-\nu} J^\mu_{\nu,0}(z) = \left(\frac{z}{2} \right)^{-\nu} J^{\mu,1}_{\nu,0}(z). \tag{6.39}$$

Additionally, if $\mu = 1$, then from (6.3) we get the classical Bessel function

$$J_\nu(z) = J^{(1)}_\nu(z) = J^1_{\nu,0}(z) = J^{1,1}_{\nu,0}(z). \tag{6.40}$$

(ii) Let $\mu = 1$, then (6.3) leads to the classical Lommel function $s_{2\lambda+\nu-1,\nu}(z)$ [Erdélyi *et al.* (1953, Vol. 2, 7.5.5 (69))]:

$$\frac{2^{2-2\lambda-\nu}}{\Gamma(\lambda)\Gamma(\lambda+\nu)} s_{2\lambda+\nu-1,\nu}(z) = J^1_{\nu,\lambda}(z) = J^{1,1}_{\nu,\lambda}(z). \tag{6.41}$$

In particular, if $\lambda = 1/2$, then from (6.3) we get the classical Struve function [Erdélyi *et al.* (1953, Vol. 2, 7.5.5 (84))]:

$$\mathbf{H}_\nu(z) = \frac{2^{1-\nu}}{\sqrt{\pi}\Gamma\left(\nu + \frac{1}{2}\right)} s_{\nu,\nu}(z) = J^1_{\nu,\frac{1}{2}}(z) = J^{1,1}_{\nu,\frac{1}{2}}(z). \tag{6.42}$$

6.6. Asymptotics for Lommel and Struve Functions

The asymptotic formula for the generalized Bessel–Maitland functions is given in Sec. 6.4 just as a corollary of Theorem 6.5. This one, for the Bessel–Maitland functions, is proved independently, regardless of the fact that it could be deduced from Theorem 6.5 (or Corollary 6.1). For $\mu = 1$, one can obtain the corresponding asymptotics for the Lommel and Struve functions, as well. As previously mentioned (see (6.11) and (6.41)), the generalized Bessel–Maitland functions (6.3) turn into Lommel functions $s_{\alpha,\nu}$ [Erdélyi *et al.* (1953, Vol. 2, 7.5.5 (69))] for parameter $\mu = 1$. This correlation (see [Marichev (1978, p. 110, (8.3)); Kiryakova (1994, pp. 352–353)]) can be rewritten as

$$s_{\alpha,\nu}(z) = 2^{\alpha-1}\Gamma\left(\frac{\alpha-\nu+1}{2}\right)\Gamma\left(\frac{\alpha+\nu+1}{2}\right) J^1_{\nu,\frac{\alpha-\nu+1}{2}}(z), \quad (6.43)$$

or, by taking $\lambda = \frac{\alpha-\nu+1}{2}$, in the form:

$$s_{\alpha,\alpha+1-2\lambda}(z) = 2^{\alpha-1}\Gamma(\lambda)\Gamma(\alpha+1-\lambda) J^1_{\alpha+1-2\lambda,\lambda}(z). \quad (6.44)$$

Additionally, if $\lambda = 1/2$, that is $\alpha = \nu$, applying (6.44) we obtain the Struve functions [Erdélyi *et al.* (1953, Vol. 2, 7.5.5 (84))]:

$$\mathbf{H}_\nu(z) = \frac{2^{1-\nu}}{\sqrt{\pi}\Gamma\left(\nu+\frac{1}{2}\right)} s_{\nu,\nu}(z) = J^1_{\nu,\frac{1}{2}}(z). \quad (6.45)$$

Then, Theorem 6.5 used with $m = 1$, or simply Corollary 6.1, provides the following corollaries.

Corollary 6.2. *The following asymptotic formula:*

$$\begin{aligned} s_{m,m+1-2\lambda}(z) &= 2^{m-1}\Gamma(\lambda)\Gamma(m+1-\lambda) J^1_{m+1-2\lambda,\lambda}(z) \\ &= [4\lambda(m+1-\lambda)]^{-1} z^{m+1}\left(1 + \theta^1_{m+1-2\lambda,\lambda}(z)\right) \end{aligned}$$
$$(6.46)$$

holds for the Lommel functions (6.44), *with*

$$\theta^1_{m+1-2\lambda,\lambda}(z) \to 0 \quad as \quad m \to \infty \quad (m \in \mathbb{N}), \quad (6.47)$$

the convergence being uniform on the compact subsets of complex z-plane.

Corollary 6.3. *The Struve functions* (6.45) *have the following asymptotic formula:*

$$H_n(z) = J^1_{n,\frac{1}{2}}(z) = \frac{2\left(\frac{z}{2}\right)^n}{\sqrt{\pi}\,\Gamma(n+\frac{1}{2})}\left(1+\theta^1_{n,\frac{1}{2}}(z)\right), \qquad (6.48)$$

with

$$\theta^1_{n,\frac{1}{2}}(z) \to 0 \quad as \quad n \to \infty \quad (n \in \mathbb{N}), \qquad (6.49)$$

the convergence being uniform on the compact subsets of complex z-plane.

Chapter 7

Mittag-Leffler Type Functions

7.1. Mittag-Leffler Functions

In 1899, Mittag-Leffler began the publication of a series of articles under the common title 'Sur la représentation analytique d'une branche uniforme d'une fonction monogène' ('On the analytic representation of a single-valued branch of a monogene function') published mainly in 'Acta Mathematica'. His research was connected with the solution of a problem of analytic continuation of complex-valued functions represented by power series. The function which he used for the solution of this problem was later named as *Mittag-Leffler function*. He defined it by a power series in the following way

$$E_\alpha(z) = \sum_{k=0}^{\infty} \frac{z^k}{\Gamma(\alpha k + 1)}, \quad z \in \mathbb{C}, \ \alpha \in \mathbb{C}, \ \Re(\alpha) > 0. \qquad (7.1)$$

The basic properties of this function were studied by Mittag-Leffler [1899, 1900a, 1900b, 1902, 1905, 1920] and Wiman [1905] as well. As it was noted by Mittag-Leffler himself, for all the values of the parameter α with $\Re(\alpha) > 0$ the series (7.1) converges in the whole complex plane, and thus it is an entire function of a complex variable z. It is well known that if $\alpha > 0$, then (7.1) gives an example of an entire function of order $\rho = 1/\alpha$ and type $\sigma = 1$.

For special values of parameter α, the function $E_\alpha(z)$ coincides with some elementary and special functions. In particular,

$$E_1(z) = \sum_{k=0}^{\infty} \frac{z^k}{\Gamma(k+1)} = \exp(z),$$

$$E_2\left(-z^2\right) = \sum_{k=0}^{\infty} \frac{(-1)^k z^{2k}}{\Gamma(2k+1)} = \cos z,$$

$$E_2\left(z^2\right) = \sum_{k=0}^{\infty} \frac{z^{2k}}{\Gamma(2k+1)} = \cosh z.$$

This function provides a simple generalization of the exponential function because of the replacement of $k! = \Gamma(k+1)$ by $\Gamma(\alpha k + 1)$ in the denominator of the power terms of the exponential series, and hence some times, it is called a *generalized exponential*.

During the first half of the 20th century, the Mittag-Leffler function remained almost unknown to the majority of scientists. Probably for the first time, an interest to this function from the application appeared due to representation in terms of this function the solution of the Abel's integral equation of the second-order made by Hille and Tamarkin (1930). A description of the most important properties of this function is present in the third volume of the Handbook on Higher Transcendental Functions of the Bateman Project [Erdélyi *et al.* (1955)]. In it, the authors have included the Mittag-Leffler functions in Chapter XVIII devoted to the so-called miscellaneous functions. The attribution of 'miscellaneous' to the Mittag-Leffler function is due to the fact that it was only later, in the 1960s, when it was recognized to belong to a more general class of higher transcendental functions, known as Fox's H-functions (see e.g. [Mathai and Saxena (1978); Kilbas and Saigo (2004); Mathai *et al.* (2010)]).

Nowadays this function and its numerous generalizations are involved in different fractional models (see e.g. the monographs listed below at the end of this section). Special role of the Mittag-Leffler function was pointed out by Kiryakova [2010a, 2010b], who included it into the class of special functions for FC. Moreover, based on the role of the Mittag-Leffler function in application, Mainardi called it *the Queen of fractional calculus* (see e.g. [Mainardi (2010)]). Ever since this exceptional role of the collection of Mittag-Leffler functions, any new exact result involving these functions seems very interesting.

First generalization of the function $E_\alpha(z)$, also called *Mittag-Leffler function*, that is given below,

$$E_{\alpha,\beta}(z) = \sum_{k=0}^{\infty} \frac{z^k}{\Gamma(\alpha k + \beta)}, \quad z \in \mathbb{C}, \ \alpha, \beta \in \mathbb{C}, \ \Re(\alpha) > 0, \quad (7.2)$$

was in fact introduced by Agarwal [1953], though some facts were earlier (almost by five decades) mentioned by Wiman in [Wiman (1905)]. A number of relationships for this function were obtained by Humbert and Agarwal [1953] using the Laplace transformation technique. This function could have been called the Agarwal function. However, Agarwal and Humbert generously left the same name as for the one-parameter Mittag-Leffler function, and that was the reason why the two-parameter function is now called the Mittag-Leffler function. The function (7.2) reduces itself to the classical Mittag-Leffler function if we put $\beta = 1$. Analogously to (7.1), the series in (7.2) is also convergent in the whole complex plane $\mathbb{C}$, and so $E_{\alpha,\beta}(z)$ is an entire function. Like the Mittag-Leffler functions $E_\alpha(z)$, the function $E_{\alpha,\beta}(z)$ is an entire function of the same order $\rho = 1/\Re(\alpha)$ [Kilbas *et al.* (2006, p. 42)]. In particular, if $\alpha > 0$, both functions have order $\rho = 1/\alpha$ and type $\sigma = 1$. They both have been studied in detail by Dzrbashjan [1960, 1966] including asymptotic formulae in different parts of the complex plane, distribution of the zeros, kernel functions of inverse Borel type integral transforms, various relations and representations. The most essential properties and applications of these entire functions, investigated by many mathematicians, can be found in the contemporary monographs of Kilbas *et al.* [2006], Gorenflo *et al.* [2014], Baleanu *et al.* [2012] and Podlubny [1999].

The results in this chapter refer to the estimations and asymptotic formulae concerning Mittag-Leffler functions and some of their generalizations. They are largely based on the author's papers [Paneva-Konovska (2010c, 2012a)].

7.2. Inequalities Related to the Mittag-Leffler Functions

In this section, we provide some useful estimates connected to the above-discussed Mittag-Leffler functions with positive α. We first take $\beta = n \in \mathbb{N}$, then we introduce the denotations

$$\theta_n(z) = E_n(z) - 1 = \sum_{k=1}^{\infty} \frac{z^k}{\Gamma(kn+1)}, \tag{7.3}$$

$$\theta_{\alpha,n}(z) = \Gamma(n)E_{\alpha,n}(z) - 1 = \Gamma(n) \sum_{k=1}^{\infty} \frac{z^k}{\Gamma(\alpha k + n)}. \tag{7.4}$$

Now, leaving α to be a positive number, we establish the statement given below.

Theorem 7.1. *Let $n \in \mathbb{N}$, $\alpha \in \mathbb{R}^+$, $z \in \mathbb{C}$ and $K \subset \mathbb{C}$ be a nonempty compact set. Then the following inequalities hold:*

$$|\theta_n(z)| \le \frac{1}{n!} \left(\exp\left(|z|\right) - 1 \right), \tag{7.5}$$

$$|\theta_{\alpha,n}(z)| \le \frac{\Gamma(n)}{\Gamma(\alpha+n)} \, \Gamma(\alpha+1) \left(E_\alpha(|z|) - 1 \right). \tag{7.6}$$

Moreover, there exists a constant $C = C(K)$, $0 < C < \infty$, such that

$$|\theta_n(z)| \le \frac{C}{n!}, \quad |\theta_{\alpha,n}(z)| \le C \frac{\Gamma(n)}{\Gamma(\alpha+n)}, \tag{7.7}$$

for all the natural numbers n and each $z \in K$.

Proof. First, let $z \in \mathbb{C}$. Then $\theta_n(z)$ can be rewritten in the following way

$$\theta_n(z) = \frac{1}{\Gamma(n+1)} \sum_{k=1}^{\infty} \frac{\Gamma(n+1)}{\Gamma(kn+1)} \, z^k = \frac{1}{n!} \sum_{k=1}^{\infty} \frac{n!}{(kn)!} \, z^k.$$

Denoting $u_k(z) = \frac{n!}{(kn)!} z^k$, we obtain the estimate $|u_k(z)| \leq \frac{|z|^k}{k!}$ for the absolute value of $u_k(z)$. Since the series $\sum_{k=1}^{\infty} \frac{|z|^k}{k!}$ converges for each $z \in \mathbb{C}$ with a sum $\exp(|z|) - 1$, then the estimate (7.5) holds in the whole complex plane.

In order to prove (7.6), we write (7.4) as follows:

$$\theta_{\alpha,n}(z) = \frac{\Gamma(n)}{\Gamma(\alpha + n)} \sum_{k=1}^{\infty} \frac{\Gamma(\alpha + n)}{\Gamma(\alpha k + n)} z^k,$$

and denoting

$$\widetilde{\gamma}_{n,k} = \frac{\Gamma(\alpha + n)}{\Gamma(\alpha k + n)}, \quad \widetilde{u}_{n,k}(z) = \widetilde{\gamma}_{n,k} z^k,$$

we obtain consecutively

$$\widetilde{\gamma}_{n,1} = 1,$$

$$0 < \widetilde{\gamma}_{n,k} = \frac{\Gamma(\alpha + 1)}{\Gamma(\alpha k + 1)} \prod_{s=1}^{n-1} \frac{(\alpha + s)}{(\alpha k + s)} \leq \frac{\Gamma(\alpha + 1)}{\Gamma(\alpha k + 1)}, \quad \text{for } k \in \mathbb{N}, \, k \neq 1,$$

$$|\widetilde{u}_{n,k}(z)| = \widetilde{\gamma}_{n,k}|z|^k \leq \frac{\Gamma(\alpha + 1)}{\Gamma(\alpha k + 1)}|z|^k, \quad \text{for } k \in \mathbb{N},$$

and therefore,

$$|\theta_{\alpha,n}(z)| \leq \frac{\Gamma(n)\Gamma(\alpha + 1)}{\Gamma(\alpha + n)} \left(\sum_{k=0}^{\infty} \frac{|z|^k}{\Gamma(\alpha k + 1)} - 1 \right)$$

which proves (7.6).

Further, for all z on the compact set K, the inequalities (7.7) immediately follow from the inequalities (7.5) and (7.6). $\qquad \square$

7.3. Generalized Mittag-Leffler Functions

Further generalization of the function $E_\alpha(z)$ was proposed by Prabhakar (1971) who generalized (7.1) by introducing a three-parameter function $E_{\alpha,\beta}^{\gamma}$ of the form:

$$E_{\alpha,\beta}^{\gamma}(z) = \sum_{k=0}^{\infty} \frac{(\gamma)_k}{\Gamma(\alpha k + \beta)} \frac{z^k}{k!}, \quad z \in \mathbb{C}, \ \alpha, \ \beta, \ \gamma \in \mathbb{C}, \ \Re(\alpha) > 0,$$

(7.8)

where $(\gamma)_k$ is the *Pochhammer symbol* [Erdélyi *et al.* (1953, Section 2.1.1)]:

$$(\gamma)_0 = 1, \quad (\gamma)_k = \gamma(\gamma + 1) \dots (\gamma + k - 1), \quad (7.9)$$

it is being understood conventionally that $(0)_0 := 1$ and assumed tacitly that the Γ-quotient exists. For $\gamma = 1$, the function $E_{\alpha,\beta}^{\gamma}$ coincides with $E_{\alpha,\beta}$, while for $\gamma = \beta = 1$, it coincides with E_{α}, i.e.

$$E_{\alpha,\beta}^{1}(z) = E_{\alpha,\beta}(z), \quad E_{\alpha,1}^{1}(z) = E_{\alpha}(z).$$

The *generalized three-parameter Mittag-Leffler function* (7.8) is an entire function of z of order $\rho = 1/\Re(\alpha)$, as it is mentioned in [Kilbas *et al.* (2006); Mathai and Haubold (2011)]. Prabhakar himself studied some properties of this function and of an integral operator containing it as a kernel function, and applied the obtained results to prove the existence and uniqueness of the solution of a corresponding integral equation. Further, some properties of $E_{\alpha,\beta}^{\gamma}(z)$, including the classical and fractional order differentiations and integrations, are proved by Kilbas *et al.* [2004]. For the various properties of the Prabhakar's Mittag-Leffler type function, see also the monograph of Kilbas *et al.* [2006], and for its recent use in the friction and generalized memory kernels and in the exact solutions of the fractional generalized Langevin equation, see, for example, Sandev *et al.* [2012, 2014], as well as Sandev *et al.* [2015]. An integral operator with such a kernel function is also studied in the space $L(a, b)$. The functions (7.8) and series in them have been recently used to express solutions of the generalized Langevin equation by Sandev *et al.* [2011], as well as in the eigenfunction expansion of the solution of two-term time-fractional equations by Bazhlekova and Dimovski [2013, 2014]. Another type of three-parameter Mittag-Leffler function (q-analog of the Mittag-Leffler function) is also considered, see, for example, Rajkovic *et al.* [2008]. Actually, the special role of the Mittag-Leffler type functions in FC has been discovered by many scientists from a different point of view. Plenty of various properties of these functions are studied by many authors, among them are [Gajić and Stanković (1976); Kilbas *et al.* (2006); Kiryakova (2000, 2010a, 2010b); Podlubny (1999); Rajkovic *et al.* (2008); Hilfer (2000)], just to name a few.

7.4. Results on the Parameters of the Generalized Mittag-Leffler Functions

Now consider the generalized (three-parameter) Mittag-Leffler functions (7.8) for indices of the kind $\beta = n$, $n = 0, 1, 2, \ldots$, namely:

$$E_{\alpha,n}^{\gamma}(z) = \sum_{k=0}^{\infty} \frac{(\gamma)_k}{\Gamma(\alpha k + n)} \frac{z^k}{k!}, \quad \alpha, \, \gamma \in \mathbb{C}, \, \Re(\alpha) > 0, \, n \in \mathbb{N}_0.$$

$$(7.10)$$

Remark 7.1. Depending on the parameters, some coefficients in (7.10) might be equal to zero. In this section, we are going to establish that there exists a number $p \in \mathbb{N}_0$ such that the representation (7.10) can be written as follows:

$$E_{\alpha,n}^{\gamma}(z) = z^p \sum_{k=p}^{\infty} \frac{(\gamma)_k}{\Gamma(\alpha k + n)} \frac{z^{k-p}}{k!}, \quad (7.11)$$

determining the respective values of p.

For this purpose, setting for convenience

$$a_k = \frac{1}{\Gamma(\alpha k + n)}, \quad b_k = (\gamma)_k, \quad c_k = \frac{a_k b_k}{k!}, \quad k = 0, 1, 2, \ldots,$$

we separately consider three main cases. Beginning with $\gamma \neq 0, -1, -2, \ldots$, we state the following lemma.

Lemma 7.1. *If $\gamma \in \mathbb{C}$, but $\gamma \notin \mathbb{Z}_0^-$, then*

(i) *$p = 0$ for $n \in \mathbb{N}$,*
(ii) *$p = 1$ for $n = 0$.*

Proof. Obviously, in the first case, $b_k \neq 0$ and $\alpha k + n$ are neither negative integer numbers nor zero. Because of that, $a_k \neq 0$ and therefore $c_k \neq 0$ for all the values of k. In the second case, since only a_0 equals zero, then $c_0 = 0$, but $c_k \neq 0$ for all the natural values of k.
□

In particular, if $\gamma = 1$, the result obtained in Lemma 7.1 refers to the two-parameter functions $E_{\alpha,n}(z)$, i.e. all the coefficients of its power representation are different from zero, probably except for c_0. Namely, the statement given below holds.

Corollary 7.1. *Let $\alpha, z \in \mathbb{C}$ and $\Re(\alpha) > 0$. Then the Mittag-Leffler functions $E_{\alpha,n}(z)$ satisfy the relation:*

$$E_{\alpha,n}(z) = z^p \sum_{k=p}^{\infty} \frac{z^{k-p}}{\Gamma(\alpha k + n)}, \tag{7.12}$$

with the respective values of p, namely:

(i) *$p = 0$ for $n \in \mathbb{N}$,*
(ii) *$p = 1$ for $n = 0$.*

Remark 7.2. Actually, in both other cases, the functions (7.10) reduce to polynomials of power $-\gamma$, and denoting $m = -\gamma$, we can rewrite representation (7.11) in the form:

$$E_{\alpha,n}^{\gamma}(z) = z^p \sum_{k=p}^{m} \frac{(\gamma)_k}{\Gamma(\alpha k + n)} \frac{z^{k-p}}{k!}. \tag{7.13}$$

Lemma 7.2. *If $\gamma \in \mathbb{Z}^-$ and $m = -\gamma$, then (7.10) can be expressed by the formula (7.13) with the following values of p:*

(i) *$p = 0$ for $n \in \mathbb{N}$,*
(ii) *$p = 1$ for $n = 0$.*

Proof. The values of the numbers a_k are the same as in the proof of Lemma 7.1. Moreover, with respect to the numbers b_k, one can write

$$\begin{aligned} b_k &= (-m)_k = -m(-m+1)\cdots(-m+k-1) \\ &= (-1)^k m(m-1)\cdots(m-k+1), \end{aligned}$$

from which $b_k \neq 0$ only for $k \leq m$ and because of that $c_k = 0$ for all $k > m$. Therefore, the values of p are the same as it is required to be proven. $\qquad\square$

Corollary 7.2. *If $\gamma \in \mathbb{Z}^-$ and $m = -\gamma$, then (7.13) can be written in the form:*

$$E_{\alpha,n}^{-m}(z) = z^p \sum_{k=p}^{m} (-1)^k \binom{m}{k} \frac{z^{k-p}}{\Gamma(\alpha k + n)}, \qquad (7.14)$$

with the corresponding values of p.

Proof. Taking into account the relation $c_k = a_k b_k / k!$ and the expression for b_k, obtained by proving Lemma 7.2, the validity of (7.14) automatically follows. $\qquad\square$

Remark 7.3. Let us mention that $(-1)^p \binom{m}{p} = (-m)_p$ when $p = 0$ or $p = 1$ and $m \in \mathbb{N}$.

Lemma 7.3. *If $\gamma = 0$, then:*

(i) $E_{\alpha,n}^0(z) = \frac{1}{\Gamma(n)}$ *for $n \in \mathbb{N}$,*

(ii) $E_{\alpha,n}^0(z) = 0$ *for $n = 0$.*

Proof. Taking in view that $b_k = 0$ for all $k \in \mathbb{N}$, the sum in (7.8) is reduced to the first addend c_0. The rest follows directly, after calculating this term. $\qquad\square$

Remark 7.4. Let us mention that if γ is a nonpositive integer, as it is seen above, the functions (7.8) reduce to polynomials. However, when $\gamma \notin \mathbb{Z}_0^-$, they are entire functions of z of order $\rho = 1/\Re(\alpha)$ and this is not hard to be verified.

The above lemmas and Remark 7.3 show that the functions $E_{\alpha,n}^{\gamma}(z)$ can be written in the following form:

$$E_{\alpha,n}^{\gamma}(z) = \frac{(\gamma)_p}{\Gamma(\alpha p + n)}\, z^p \left(1 + \theta_{\alpha,n}^{\gamma}(z)\right), \qquad (7.15)$$

with

$$\theta_{\alpha,n}^{\gamma}(z) = \sum_{k=p+1}^{\infty} \frac{(\gamma)_k}{(\gamma)_p} \frac{\Gamma(\alpha p + n)}{\Gamma(\alpha k + n)} \frac{z^{k-p}}{k!} \quad \text{for } \gamma \in \mathbb{C}\backslash\mathbb{Z}_0^-, \qquad (7.16)$$

and respectively, for $\gamma = -m$, $m \in \mathbb{N}$,

$$\theta_{\alpha,n}^{-m}(z) = \sum_{k=p+1}^{m} \frac{(-m)_k}{(-m)_p} \frac{\Gamma(\alpha p + n)}{\Gamma(\alpha k + n)} \frac{z^{k-p}}{k!}$$

$$= \sum_{k=p+1}^{m} \frac{(-1)^{k-p}\binom{m}{k}}{\binom{m}{p}} \frac{\Gamma(\alpha p + n)}{\Gamma(\alpha k + n)} z^{k-p}. \qquad (7.17)$$

Remark 7.5. In the above representations (7.15)–(7.17), γ is different from zero, and the parameter p is determined by means of the previous lemmas. More precisely, $p = 0$ for all the natural values of n and $p = 1$ for $n = 0$. The functions $E_{\alpha,n}^0$ have the simplest form, given by Lemma 7.3, and moreover $E_{\alpha,0}^0 = 0$.

7.5. Inequalities Related to the Generalized Mittag-Leffler Functions

In this section, our aim is to estimate the entire functions $\theta_{\alpha,n}^{\gamma}(z)$ for nonnegative integer values of n, i.e. $n \in \mathbb{N}_0$, $\alpha, \gamma \in \mathbb{C}$, $\Re(\alpha) > 0$, $z \in \mathbb{C}$. To this end, we transform the expressions in the equalities (7.16) and (7.17) to the following forms:

$$\theta_{\alpha,n}^{\gamma}(z) = \frac{\Gamma(\alpha p + n)}{\Gamma(\alpha(p+1) + n)} \sum_{k=p+1}^{\infty} \gamma_{n,k} \frac{(\gamma)_k}{(\gamma)_p} \frac{z^{k-p}}{k!}, \qquad (7.18)$$

for $\gamma \in \mathbb{C}\backslash\mathbb{Z}_0^-$, and respectively,

$$\theta_{\alpha,n}^{-m}(z) = \frac{\Gamma(\alpha p + n)}{\Gamma(\alpha(p+1) + n)} \sum_{k=p+1}^{m} (-1)^{k-p}\, \gamma_{n,k} \frac{\binom{m}{k}}{\binom{m}{p}} z^{k-p}, \qquad (7.19)$$

for $m \in \mathbb{N}$, with

$$\gamma_{n,k} = \frac{\Gamma(\alpha(p+1)+n)}{\Gamma(\alpha k+n)}. \qquad (7.20)$$

Theorem 7.2. *Let* $\gamma \in \mathbb{C} \backslash \mathbb{Z}_0^-$, $n \in \mathbb{N}_0$ *and* $K \subset \mathbb{C}$ *be a nonempty compact set. Then there exists an entire function* τ *such that*

$$\left| \theta_{\alpha,n}^{\gamma}(z) \right| \leq \frac{|\Gamma(\alpha p+n)|}{|\Gamma(\alpha(p+1)+n)|} \tau(|z|; \alpha, \gamma), \qquad (7.21)$$

for all the values of $z \in \mathbb{C}$. *Moreover, a constant* $0 < C = C(K) < \infty$ *exists such that*

$$\left| \theta_{\alpha,n}^{\gamma}(z) \right| \leq C \frac{|\Gamma(\alpha p+n)|}{|\Gamma(\alpha(p+1)+n)|}, \qquad (7.22)$$

for all the nonnegative integers n *and each* $z \in K$.

Proof. To find such a function τ and to prove the inequality (7.21), we estimate the function (7.18) beginning with the values of (7.20). Since

$$\gamma_{0,\,k} = \frac{\Gamma(\alpha(p+1))}{\Gamma(\alpha k)},$$

$$\gamma_{n,k} = \frac{\Gamma(\alpha(p+1))}{\Gamma(\alpha k)} \frac{(\alpha(p+1))}{(\alpha k)} \cdots \frac{(\alpha(p+1)+n-1)}{(\alpha k)+n-1}, \quad \text{for } n \in \mathbb{N},$$

and due to the following inequality

$$\frac{|\alpha(p+1)|}{|\alpha k|} \cdots \frac{|\alpha(p+1)+n-1|}{|\alpha k+n-1|} \leq 1,$$

we obtain that

$$|\gamma_{n,k}| \leq \frac{|\Gamma(\alpha(p+1))|}{|\Gamma(\alpha k)|},$$

for all the possible values of n and k. Finally, the proof of the inequality (7.21) ends, by taking

$$\tau(z; \alpha, \gamma) = \sum_{k=p+1}^{\infty} \frac{|\Gamma(\alpha(p+1))|}{|\Gamma(\alpha k)|} \frac{|(\gamma)_k|}{|(\gamma)_p|} \frac{z^{k-p}}{k!},$$

cf. with (7.21). Since τ is an entire function, as well as in view of the just proven inequality, the validity of (7.22) immediately follows.

$\square$

Theorem 7.3. *Let $n \in \mathbb{N}_0$, $\gamma \in \mathbb{Z}^-$, $m = -\gamma$ (except for the case of both $n = 0$ and $\gamma = -1$, discussed in Remark 7.6), and let $K \subset \mathbb{C}$ be a nonempty compact set. Then there exists an entire function t such that*

$$\left|\theta_{\alpha,n}^{-m}(z)\right| \leq \frac{|\Gamma(\alpha p + n)|}{|\Gamma(\alpha(p+1) + n)|} \, t(|z|; \alpha, \gamma), \qquad (7.23)$$

for all $z \in \mathbb{C}$. Moreover, a constant $0 < C = C(K) < \infty$ exists such that

$$\left|\theta_{\alpha,n}^{-m}(z)\right| \leq C \frac{|\Gamma(\alpha p + n)|}{|\Gamma(\alpha(p+1) + n)|}, \qquad (7.24)$$

for all the nonnegative integers n and each $z \in K$.

Proof. Considering (7.19), following the evaluations in the proof of Theorem 7.2 and denoting

$$t(z; \alpha, \gamma) = \sum_{k=p+1}^{m} \frac{|\Gamma(\alpha(p+1))|}{|\Gamma(\alpha k)|} \frac{\binom{m}{k}}{\binom{m}{p}} z^{k-p},$$

we complete the proof through analogy with the one of Theorem 7.2.

$\square$

Remark 7.6. Consideration of the simplest case, both $n = 0$ and $\gamma = -1$, leads to the result

$$E_{\alpha,0}^{-1}(z) = -\frac{z}{\Gamma(\alpha)}, \qquad \theta_{\alpha,0}^{-1}(z) \equiv 0. \qquad (7.25)$$

7.6. Asymptotic Formulae with Respect to the Index n

Some asymptotic formulae for 'large' values of the indices that are used in the further exposition are provided in this section.

Theorem 7.4. *Let* $\alpha > 0$ *and* $n \in \mathbb{N}$, *then the Mittag-Leffler functions* E_n *and* $E_{\alpha,n}$ *have the following asymptotic formulae:*

$$E_n(z) = 1 + \theta_n(z), \quad \theta_n(z) \to 0 \quad as\ n \to \infty, \qquad (7.26)$$

$$E_{\alpha,n}(z) = \frac{1}{\Gamma(n)}\ (1 + \theta_{\alpha,n}(z)), \quad \theta_{\alpha,n}(z) \to 0 \quad as\ n \to \infty, \qquad (7.27)$$

for $z \in \mathbb{C}$. *The functions* $\theta_n(z)$ *and* $\theta_{\alpha,n}(z)$ *are holomorphic in the complex plane* $\mathbb{C}$ *and the convergence is uniform on the compact subsets of the complex plane* $\mathbb{C}$ *as well, and*

$$\theta_n = O\left(\frac{1}{n!}\right), \quad \theta_{\alpha,n} = O\left(\frac{1}{n^\alpha}\right). \qquad (7.28)$$

Proof. The identities (7.26) and (7.27) are automatically obtained due to Eqs. (7.1)–(7.4). The holomorphy of $\theta_n(z)$ and $\theta_{\alpha,n}(z)$ follows from the holomorphy of $E_n(z)$ and $E_{\alpha,n}(z)$ on the whole complex plane and the equalities (7.26) and (7.27). The rest immediately follows in accordance with Theorem 7.1 and Remark 6.5 case (iii).

$\square$

Theorem 7.5. *Let* $z, \alpha, \gamma \in \mathbb{C}, n \in \mathbb{N}_0, \gamma \neq 0, \Re(\alpha) > 0$ *and* $\theta_{\alpha,n}^\gamma$ *be given by the formulae* (7.18)–(7.20). *Then the generalized Mittag-Leffler functions* (7.10) *have the following asymptotic formulae:*

$$E_{\alpha,n}^\gamma(z) = \frac{(\gamma)_p}{\Gamma(\alpha p + n)} z^p \left(1 + \theta_{\alpha,n}^\gamma(z)\right) \quad and$$

$$\theta_{\alpha,n}^\gamma(z) \to 0 \quad as\ n \to \infty, \qquad (7.29)$$

with the corresponding p, *depending on* γ. *Moreover, on the compact subsets of the complex plane* $\mathbb{C}$, *the convergence is uniform and*

$$\theta_{\alpha,n}^\gamma(z) = O\left(\frac{1}{n^{\Re(\alpha)}}\right) \quad (n \in \mathbb{N}). \qquad (7.30)$$

Proof. Using the formula (7.15), Theorems 7.2 and 7.3 and the property of Γ-functions quotient (Remark 6.5 case (iii)), the proof is evident. $\square$

In conclusion, note that the case $\gamma = 1$ gives analogous results related to the classical Mittag-Leffler functions $E_{\alpha,n}$, i.e. all the results obtained in Secs. 7.5 and 7.6 are, in particular, valid for $E_{\alpha,n}$ as well. Additionally, if the parameter α is positive, the results in Sec. 7.2 and Theorem 7.4 follow. So, some useful remarks can be made.

Remark 7.7. Note that for $\gamma = 1$ the relation (7.30) refers to the function $\theta_{\alpha,n}$, given in (7.4). Additionally, if $\alpha > 0$, then (7.30) coincides with the second of the formulae (7.28).

Remark 7.8. In view of Corollary 7.1, the relation (7.29) leads to the value $p = 1$ in the corresponding representation concerning the Mittag-Leffler functions $E_{\alpha,0}(z) = E_{\alpha,0}^1(z)$.

Remark 7.9. Taking into account both Lemma 7.3 and Remark 7.6, the following relations hold:

$$\theta_{\alpha,n}^0(z) = 0 \quad (n \in \mathbb{N}), \quad \theta_{\alpha,0}^{-1}(z) = 0, \quad E_{\alpha,0}^0(z) = 0. \tag{7.31}$$

Remark 7.10. According to the asymptotic formulae (7.26) and (7.29), it follows that there exists a natural number M such that the functions $E_n(z)$, $E_{\alpha,n}(z)$ and $E_{\alpha,n}^\gamma(z)$ have no zeros at all for $n > M$, possibly except for the zero.

Remark 7.11. Each functions $E_n(z)$, $E_{\alpha,n}(z)$ and $E_{\alpha,n}^\gamma(z)$ $(n \in \mathbb{N})$, being an entire function, not identically zero, has no more than a finite number of zeros in the closed and bounded set $|z| \leq R$. Moreover, because of Remark 7.10, no more than a finite number of these functions have some zeros, possibly except for the origin.

Chapter 8

Latest Generalizations of Both the Bessel and Mittag-Leffler Type Functions

8.1. Multi-index Mittag-Leffler Functions

Although the Mittag-Leffler functions have remained unknown for a long time, nowadays various generalizations have appeared in relation to their applications as solutions of fractional order equations and systems those model various phenomena and numerical algorithms. At the end of the 20th century, a class of special functions of Mittag-Leffler type those are multi-index analogs of $E_{\alpha,\beta}(z)$ has been introduced and studied, as well, by Yakubovich and Luchko [1994], Luchko and Gorenflo [1998], Luchko [1999] and Kiryakova [1999, 2000, 2008, 2010a, 2010b].

Following denotations from Kiryakova [1999], the indices $\alpha :=$ $1/\rho$ and $\beta := \mu$ are replaced by two sets of multi-indices, or vector indices, namely $\alpha \to (1/\rho_1, 1/\rho_2, \ldots, 1/\rho_m)$ and $\beta \to (\mu_1, \mu_2, \ldots, \mu_m)$ of dimension $m = 1, 2, 3, \ldots$.

Definition 8.1. Let $m > 1$ be an integer, and $\rho_1, \ldots, \rho_m > 0$ and $\mu_1, \ldots, \mu_m$ be arbitrary real (complex) numbers. By means of these 'multi-indices', the *multi-index (vector) Mittag-Leffler functions* are defined as

$$
E_{\left(\frac{1}{\rho_i}\right),(\mu_i)}(z) = E^m_{\left(\frac{1}{\rho_i}\right),(\mu_i)}(z) = \sum_{k=0}^{\infty} \frac{z^k}{\Gamma\left(\frac{k}{\rho_1} + \mu_1\right) \ldots \Gamma\left(\frac{k}{\rho_m} + \mu_m\right)}.
$$
(8.1)

This interesting class of functions appeared for the first time as explicit solutions to Cauchy type problems for a fractional differential equation of fractional multi-order in the monograph [Yakubovich and Luchko (1994)]. The theory of multi-index Mittag-Leffler functions was developed in detail in a series of papers by Kiryakova *et al.* Namely, in the paper [Kiryakova (2000, 2010a, 2010b)], it is proved that the multi-index Mittag-Leffler functions (8.1) are entire functions of order ρ and type σ, with

$$\frac{1}{\rho} = \frac{1}{\rho_1} + \cdots + \frac{1}{\rho_m}, \quad \sigma = \left(\frac{\rho_1}{\rho}\right)^{\frac{\rho}{\rho_1}} \cdots \left(\frac{\rho_m}{\rho}\right)^{\frac{\rho}{\rho_m}}. \tag{8.2}$$

There, and also in the paper [Kiryakova (2010a, 2010b)], a series of other basic properties, integral and differential formulae, etc. for (8.1) can be found. For the applications of this class of special functions in the solutions of fractional order equations and models, see e.g. in Kiryakova and Luchko [2010]. The survey by Kilbas *et al.* [2013] describes the historical development of the theory of these multi-index ($2m$-parametric) Mittag-Leffler functions as a subclass of the Wright generalized hypergeometric functions $_p\Psi_q(z)$. The method of Mellin–Barnes type integral representations allows these authors to extend the considered functions and to study them in the case of arbitrary values of parameters (for complex parameters see e.g. the works [Kilbas and Koroleva (2005, 2006)]).

The multi-index Mittag-Leffler functions (8.1) can also be seen as 'fractional indices' analogs of the Bessel and hyper-Bessel functions (*cf.* with (6.7)):

$$J^{(m-1)}_{\nu_1,\ldots,\nu_{m-1}}(z) = \left(\frac{z}{m}\right)^{\sum_{i=1}^{m-1} \nu_i} E^m_{(1,\ldots,1),(\nu_1+1,\ldots,\nu_{m-1}+1,1)}\left(-\left(\frac{z}{m}\right)^m\right), \tag{8.3}$$

and generalizations of a long list of special functions used in mathematical physics, FC and as a solution of various mathematical models (see, for example, the surveys by Kiryakova [2008; 2010a], Kiryakova and Luchko [2010]).

8.2. Results Related to the Parameters of the Multi-index Mittag-Leffler Functions

Let us fix $1 \leq i_0 \leq m$ and, for the sake of brevity, set

$$\alpha_i = \frac{1}{\rho_i} \quad (\alpha_i > 0, \ i = 1, 2, \ldots, m),$$

$$\prod_{i=1}^{m}{}' \Gamma(\alpha_i k + \mu_i) = \prod_{i=1, i \neq i_0}^{m} \Gamma(\alpha_i k + \mu_i), \tag{8.4}$$

$$(\mu_{i_0}(n)) = (\mu_1, \ldots, \mu_{i_0-1}, n, \mu_{i_0+1}, \ldots, \mu_m) = (\mu_i)|_{\mu_{i_0}=n}.$$

Consider now the multi-index Mittag-Leffler functions with indices of the kind (α_i), $(\mu_{i_0}(n))$, i.e.

$$E_{(\alpha_i),(\mu_{i_0}(n))}(z) = \sum_{k=0}^{\infty} \frac{z^k}{\Gamma(\alpha_{i_0} k + n) \prod_{i=1}^{m}{}' \Gamma(\alpha_i k + \mu_i)} \quad \text{for } n \in \mathbb{N}_0.$$
$$\tag{8.5}$$

Remark 8.1. Depending on α_i and μ_i ($i = 1, 2, \ldots, m$), some coefficients in (8.5) may be equal to zero, that is possible only when some of the numbers μ_i are nonpositive real numbers, but no more than a finite number of a coefficients might be zero. More precisely, if $\mu_{i_q} \in \mathbb{R}_0^-$ for $q = 1, 2, \ldots, M$ ($1 \leq M \leq m$), then all the coefficients with $k > \max_{1 \leq q \leq M} \left(-\frac{\mu_{i_q}}{\alpha_{i_q}}\right)$, $k \in \mathbb{N}_0$, are different from zero.

That is, there exist numbers $p \in \mathbb{N}_0$ and $s \in \mathbb{N}$, such that the identity (8.5) can be written in the form:

$$E_{(\alpha_i),(\mu_{i_0}(n))}(z)$$

$$= \frac{z^p}{\Gamma(\alpha_{i_0} p + n) \prod_{i=1}^{m}{}' \Gamma(\alpha_i p + \mu_i)}$$

$$+ \sum_{k=p+s}^{\infty} \frac{z^k}{\Gamma(\alpha_{i_0} k + n) \prod_{i=1}^{m}{}' \Gamma(\alpha_i k + \mu_i)} \quad \text{for } n \in \mathbb{N}_0. \tag{8.6}$$

Remark 8.2. Recall that we use the notations introduced in Remark 6.1. Further, set

$$a_{i_0}(k) = \frac{1}{\Gamma(\alpha_{i_0}k + n)}, \quad a_i(k) = \frac{1}{\Gamma(\alpha_i k + \mu_i)} \quad \text{for } i \neq i_0,$$

$$A_k = \prod_{i=1, \, i \neq i_0}^{m} a_i(k), \quad k = 0, 1, 2, \ldots. \tag{8.7}$$

To determine the corresponding values of p and s in (8.6), we consider three main cases. For this purpose, along with the denotations, introduced in Remarks 6.1 and 8.2, we also use

$$c_k = A_k \, a_{i_0}(k), \quad k = 0, 1, 2, \ldots. \tag{8.8}$$

Lemma 8.1. *If $\mu_i \in \mathbb{C}$ for $i \neq i_0$ but μ_i are neither negative real numbers nor zero and n is a nonnegative integer, then*

(i) $p = 0$ *and* $s = 1$ *for* $n \in \mathbb{N}$,
(ii) $p = 1$ *and* $s = 1$ *for* $n = 0$.

Proof. Obviously, since either $\mu_i > 0$ or $\mu_i \notin \mathbb{R}$, then the numbers $\alpha_i k + \mu_i \notin \mathbb{Z}_0^-$ for $i \neq i_0$ and $k = 0, 1, 2, \ldots$. Moreover, in the first case, $\alpha_{i_0}k + n > 0$. For these reasons, $A_k \neq 0$, $a_{i_0}(k) \neq 0$, and therefore $c_k \neq 0$, for all the values of k.

In the second case, since $n = 0$, then $a_{i_0}(k) = 0$ for $k = 0$, $a_{i_0}(k) > 0$ for $k = 1, 2, \ldots$, and $A_k \neq 0$ because of the same reasons like in the first case, that leads to the statements $c_0 = 0$, $c_k \neq 0$ for $k = 1, 2, \ldots$. □

Lemma 8.2. *If $i \neq i_0$, $\mu_i \in \mathbb{C} \backslash \mathbb{Z}_0^-$, at least one of $\mu_i \in \mathbb{R}^- \backslash \mathbb{Z}^-$, and n is a nonnegative integer, then there exist positive integers s_0 and l_0 such that*

(i) $p = 0$ *and* $s = s_0$ *for* $n \in \mathbb{N}$,
(ii) $p = s_0$ *and* $s = l_0$ *for* $n = 0$.

Proof. In the first case, $\alpha_{i_0}k + n > 0$ for all the values of k and therefore $a_{i_0}(k) > 0$ for $k = 0, 1, 2, \ldots$. On the other hand, for $i \neq i_0$, $A_0 \neq 0$ and then $p = 0$. However, it is possible for some numbers $\alpha_i k + \mu_i$ to belong to $\mathbb{Z}_0^-$ and then the corresponding $c_k = 0$. Therefore, in view of Remark 8.1, there exists $s_0 \in \mathbb{N}$ with $A_{s_0} \neq 0$ and because of that $c_{s_0} \neq 0$ and $s = s_0$.

In the second case, since $n = 0$, then $a_{i_0}(k) = 0$ for $k = 0$, $a_{i_0}(k) > 0$ for $k = 1, 2, \ldots$, and A_k have the same values like in the first case. Therefore, there exists a natural number l_0 such that $A_{s_0} \neq 0$, $A_{s_0+l_0} \neq 0$ but $A_k = 0$ for $s_0 < k < s_0 + l_0$, that leads to the statements $c_0 = 0$, $c_k = 0$ for $0 < k < s_0$ and $s_0 < k < s_0 + l_0$, and $c_k \neq 0$ for $k = s_0$ and $k = s_0 + l_0$ as well. Thus, $p = s_0$ and $s = l_0$. $\qquad \square$

Lemma 8.3. *If $i \neq i_0$, $\mu_i \in \mathbb{C}$, at least one of $\mu_i \in \mathbb{Z}_0^-$, and n is a nonnegative integer, then there exist positive integers p_0 and s_0 such that $p = p_0$ and $s = s_0$.*

Proof. Considering the case $n \in \mathbb{N}$, we obtain $a_{i_0}(k) > 0$ for $k = 0, 1, 2, \ldots$. On the other hand, for $i \neq i_0$, $A_0 = 0$ and it is possible for some of the numbers $\alpha_i k + \mu_i$ to belong to $\mathbb{Z}_0^-$. Therefore, there exist numbers $p_0, s_0 \in \mathbb{N}$, such that $A_{p_0} \neq 0$, $A_{p_0+s_0} \neq 0$ and $A_k = 0$ for $p_0 < k < p_0 + s_0$. Because of that, $c_{p_0} \neq 0$, $c_{p_0+s_0} \neq 0$ and $p = p_0$, $s = s_0$.

In the other case, since $n = 0$, then $a_{i_0}(k) = 0$ for $k = 0$ and $a_{i_0}(k) > 0$ for $k = 1, 2, \ldots$. Moreover, A_k have the same values like in the case $n \in \mathbb{N}$. Therefore, with the same p_0 and s_0, $c_{p_0} \neq 0$, $c_{p_0+s_0} \neq 0$ and $c_k = 0$ for $p_0 < k < p_0 + s_0$, and because of that, again $p = p_0$ and $s = s_0$. $\qquad \square$

8.3. Inequalities and Asymptotic Formulae for 'Large Values' of the Parameters μ

First, we deal with the functions (8.5). In this section, we prove an asymptotic formula for 'large' values of the index $\mu_{i_0} = n$.

Let the parameters p and s ($p \geq 0$, $s \geq 1$) be determined by means of the preceding lemmas and

$$\theta_n(z) = \sum_{k=p+s}^{\infty} \frac{\Gamma(\alpha_{i_0} p + n) \prod_{i=1}^{m} {}'\Gamma(\alpha_i p + \mu_i)}{\Gamma(\alpha_{i_0} k + n) \prod_{i=1}^{m} {}'\Gamma(\alpha_i k + \mu_i)} z^{k-p}. \tag{8.9}$$

Further, our aim is to estimate the functions $\theta_n(z)$, proving some inequalities in the complex plane $\mathbb{C}$ as well as on its compact subsets. To this end, we transform the expression in the equality (8.9) that leads to the identity

$$\theta_n(z) = \frac{\Gamma(\alpha_{i_0}p + n) \prod_{i=1}^{m}{}'\Gamma(\alpha_i p + \mu_i)}{\Gamma(\alpha_{i_0}(p + s) + n)}$$

$$\times \sum_{k=p+s}^{\infty} \frac{\gamma_{n,k}}{\prod_{i=1}^{m}{}'\Gamma(\alpha_i k + \mu_i)} z^{k-p}, \qquad (8.10)$$

with

$$\gamma_{n,k} = \frac{\Gamma(\alpha_{i_0}(p + s) + n)}{\Gamma(\alpha_{i_0} k + n)}.$$

We remind, for using in the further considerations, that there exists a number $1 < \alpha_0 < 2$ such that Euler's gamma function $\Gamma(\alpha)$ increases in (α_0, ∞) and decreases in $(0, \alpha_0)$, and on the set $\mathbb{R}^+$ of the positive real numbers, $\Gamma(\alpha)$ has its absolute minimum in the point α_0, that is $\Gamma(\alpha_0) = \min_{\alpha \in \mathbb{R}^+} \Gamma(\alpha)$ (see Remark 6.5 case (i)).

Theorem 8.1. *Let $z \in \mathbb{C}$ and $K \subset \mathbb{C}$ be a nonempty compact set. Then there exists an entire function $\varphi_{(\alpha_i),(\mu_i)}(z)$ such that for all the numbers $n \in \mathbb{N}_0$ and corresponding $p \in \mathbb{N}_0$ and $s \in \mathbb{N}$:*

$$|\theta_n(z)| \leq \frac{\Gamma(\alpha_{i_0}p + n)}{\Gamma(\alpha_{i_0}(p + s) + n)} \times \varphi_{(\alpha_i),(\mu_i)}(|z|), \qquad (8.11)$$

and moreover, there exists a constant $C = C(K)$, $0 < C < \infty$, such that

$$|\theta_n(z)| \leq C \frac{\Gamma(\alpha_{i_0}p + n)}{\Gamma(\alpha_{i_0}(p + s) + n)} \qquad (8.12)$$

for each $z \in K$.

Proof. First, we consider the case $n \geq 2$ that ensures $0 < \gamma_{n,k} \leq 1$ for all the values of k. Now, let $n = 0$ or $n = 1$. Then because of the convergence of the sequence $\{\gamma_{n,k}\}_{k=p+s}^{\infty}$, it is bounded and therefore, there exists a constant $\tilde{A}$ such that $|\gamma_{n,k}| \leq \tilde{A}$ for all the values of k.

Further, the proof of the estimates (8.11) in the whole complex plane ends taking

$$\varphi_{(\alpha_i),(\mu_i)}(z) = A \prod_{i=1}^{m}{}' |\Gamma(\alpha_i p + \mu_i)| \sum_{k=p+s}^{\infty} \frac{z^{k-p}}{\prod_{i=1}^{m}{}'|\Gamma(\alpha_i k + \mu_i)|},$$

with the constant $A = 1$ for $n \geq 2$ and $A = \tilde{A}$ for $n = 0$ and $n = 1$.

Eventually, for all $z \in K$, the inequalities (8.12) follow immediately from the inequality (8.11). □

Furthermore, an asymptotic formula for 'large' values of the index n is valid as follows.

Theorem 8.2. *Let K be a nonempty compact subset of the complex plane $\mathbb{C}$, $z \in \mathbb{C}$, $n \in \mathbb{N}_0$ and let θ_n be the functions defined by (8.9), with the corresponding $p \in \mathbb{N}_0$ and $s \in \mathbb{N}$. Then the multi-index Mittag-Leffler functions (8.5) satisfy the following asymptotic formula:*

$$E_{(\alpha_i),(\mu_{i_0}(n))}(z) = \frac{z^p}{\Gamma(\alpha_{i_0}p + n)\prod_{i=1}^{m}{}'\Gamma(\alpha_i p + \mu_i)}(1 + \theta_n(z)),$$

$$\theta_n(z) \to 0 \quad as \quad n \to \infty. \tag{8.13}$$

The functions $\theta_n(z)$ are holomorphic for $z \in \mathbb{C}$, the convergence is uniform on the nonempty compact set K of the complex plane $\mathbb{C}$ and moreover,

$$\theta_n(z) = O\left(\frac{\Gamma(\alpha_{i_0}p + n)}{\Gamma(\alpha_{i_0}(p+s) + n)}\right) = O\left(\frac{1}{n^{s\alpha_{i_0}}}\right), \quad z \in K. \tag{8.14}$$

Proof. The identity (8.13) automatically holds due to (8.5), (8.6), (8.9) and (8.10). The holomorphy of θ_n follows from the holomorphy of $E_{(\alpha_i),(\mu_{i_0}(n))}$ in the whole complex plane and the equality (8.13). The rest follows immediately from Theorem 8.1. □

Remark 8.3. Note that according to the asymptotic formula (8.13), it follows that there exists a positive integer M such that the functions (8.5) have no zeros for $n > M$, possibly except at the origin.

Remark 8.4. Each function in (8.5) ($n \in \mathbb{N}_0$), being an entire function not identically zero, has no more than a finite number of zeros in the closed and bounded set $|z| \leq R$. Moreover, because of Remark 8.3, no more than a finite number of these functions have some zeros, possibly except at the origin.

Let $z, \nu_i \in \mathbb{C}, \Re(\nu_i + 1) > 0$ ($i = 1, \ldots, m$) and $J_{\left(\nu_{i_0}(n)\right)}$ ($n = 0, 1, \ldots$) be the functions (6.20), and additionally

$$(\widetilde{\nu}_{i_0}(n+1))$$
$$= (\widetilde{\nu}_1, \ldots, \widetilde{\nu}_{i_0-1}, n+1, \widetilde{\nu}_{i_0+1}, \ldots, \widetilde{\nu}_m, \widetilde{\nu}_{m+1}) = (\widetilde{\nu}_i)|_{\widetilde{\nu}_{i_0}=n+1},$$
$$\nu_{m+1} = 0, \quad \widetilde{\nu}_i = \nu_i + 1 \quad (i = 1, 2, \ldots, m+1). \tag{8.15}$$

Then the formula (8.3) can be adapted to the function (6.20) taking the form:

$$J^{(m)}_{\left(\nu_{i_0}(n)\right)}(z) = A(z; m, n) E^{m+1}_{(1,\ldots,1),\left(\widetilde{\nu}_{i_0}(n+1)\right)} \left(-\left(\frac{z}{m+1}\right)^{m+1}\right), \tag{8.16}$$

with $A(z; m, n)$ defined by (6.21).

Now, we are going to consider the functions $E^{m+1}_{(1,\ldots,1),\left(\widetilde{\nu}_{i_0}(n+1)\right)}$, present them of the form (8.13) and find an upper estimate for the corresponding $|\vartheta_n|$. Then we apply the determined estimation to (8.16). Recall that, according to Remark 6.4, the coefficients of the functions mentioned above are all different from zero. We begin our considerations with $E^{m+1}_{(1,\ldots,1),\left(\widetilde{\nu}_{i_0}(n+1)\right)}(z)$. In view of the specific features of the discussed functions and using their explicit forms, we can write (*cf.* with (8.9) and (8.13))

$$E^{m+1}_{(1,\ldots,1),\left(\widetilde{\nu}_{i_0}(n+1)\right)}(z) = \frac{\left(1 + \widetilde{\vartheta}_n(z)\right)}{\Gamma(n+1)\prod_{i=1}^{m}{}'\Gamma(\nu_i + 1)} \tag{8.17}$$

with

$$\widetilde{\vartheta}_n(z) = \frac{1}{(n+1)} \sum_{k=1}^{\infty} \frac{\prod_{i=1}^{m}{}'\Gamma(\nu_i + 1)}{(n+2)_{k-1}\prod_{i=1}^{m}{}'\Gamma(k + \nu_i + 1)} \frac{z^k}{\Gamma(k+1)}. \tag{8.18}$$

The Theorem given below, referring to (8.18), can be easily verified.

Theorem 8.3. *Let $\widetilde{\vartheta}_n$ be the function defined by* (8.18). *Then the following inequality holds:*

$$|\widetilde{\vartheta}_n(z)| \leq \frac{|z|\exp(|z|)}{(n+1)\prod_{i=1}^{m}{}'|(\nu_i+1)|}, \tag{8.19}$$

for all the values of the variable z in the complex plane $\mathbb{C}$.

Proof. So, transforming the expression on the right-hand side of the identity (8.18), we obtain consecutively:

$$|\widetilde{\vartheta}_n(z)| \leq \frac{1}{(n+1)} \sum_{k=1}^{\infty} \frac{1}{\prod_{i=1}^{m}{}'|(\nu_i+1)_k|} \frac{|z|^k}{\Gamma(k+1)}$$

$$\leq \frac{|z|}{(n+1)\prod_{i=1}^{m}{}'|(\nu_i+1)|}$$

$$\times \left(1 + \sum_{k=2}^{\infty} \frac{1}{\prod_{i=1}^{m}{}'|(\nu_i+2)_{k-1}|} \frac{|z|^{k-1}}{\Gamma(k+1)}\right)$$

$$\leq \frac{|z|}{(n+1)\prod_{i=1}^{m}{}'|(\nu_i+1)|} \left(1 + \sum_{k=2}^{\infty} \frac{|z|^{k-1}}{\Gamma(k)}\right),$$

that confirms the validity of inequality (8.19). $\qquad\square$

Finally, the denotation

$$\vartheta_n(z) = \widetilde{\vartheta}_n\left(-\left(\frac{z}{m+1}\right)^{m+1}\right) \tag{8.20}$$

along with (8.17) lead to the relationship:

$$E_{(1,\dots,1),(\widetilde{\nu}_{i_0}(n+1))}^{m+1}\left(-\left(\frac{z}{m+1}\right)^{m+1}\right) = \frac{(1+\vartheta_n(z))}{\Gamma(n+1)\prod_{i=1}^{m}{}'\Gamma(\nu_i+1)}. \tag{8.21}$$

In conclusion, let us point out that combining the last several results allows Theorem 6.3 to be proved. Actually, the last formula has a crucial role for the proof. More precisely, the following remark can be made.

Remark 8.5. The validity of (6.32) and (6.33) follows immediately, taking into account together (8.16), (8.19) and (8.21). Eventually, the estimate (6.34) is an elementary consequence of the inequality (6.33).

8.4. Definition and Basic Properties of the Multi-index (3m-parametric) Mittag-Leffler Functions

Recently, in the paper [Paneva-Konovska (2011)] and also in the paper [Paneva-Konovska (2013b)], which are the basis of this chapter, the $3m$-parametric Mittag-Leffler functions, those generalize both Prabhakar functions (7.8) and the multi-index Mittag-Leffler functions (8.1) with $2m$ parameters, have been introduced and studied.

Definition 8.2. Let $m > 1$ be an integer and consider parameters α_i, β_i, $\gamma_i \in \mathbb{C}$, $\Re(\alpha_i) > 0$ for all $i = 1, 2, \ldots, m$. By means of the multi-indices $(\alpha_i), (\beta_i)$ and (γ_i), we introduce the so-called $3m$-*parametric* (*multi-index*) *Mittag-Leffler functions*, namely:

$$E_{(\alpha_i),(\beta_i)}^{(\gamma_i),m}(z) = \sum_{k=0}^{\infty} \frac{(\gamma_1)_k \cdots (\gamma_m)_k}{\Gamma(\alpha_1 k + \beta_1) \ldots \Gamma(\alpha_m k + \beta_m)} \frac{z^k}{(k!)^m}. \qquad (8.22)$$

In Paneva-Konovska (2011), we proved the following Theorem, which gives the order and type of (8.22).

Theorem 8.4. *If none of the parameters* $\gamma_1, \ldots, \gamma_m$ *is a negative integer or zero, then the multi-index Mittag-Leffler function* (8.22) *is an entire function of order* ρ *and type* σ *with*

$$\frac{1}{\rho} = \Re(\alpha_1) + \cdots + \Re(\alpha_m), \qquad (8.23)$$

respectively

$$\frac{1}{\sigma} = |(\rho\alpha_1)^{\rho\alpha_1}| \cdots |(\rho\alpha_m)^{\rho\alpha_m}|. \qquad (8.24)$$

Moreover, for each positive ε *the asymptotic estimate*

$$\left| E_{(\alpha_i),(\beta_i)}^{(\gamma_i),m}(z) \right| < \exp((\sigma + \varepsilon)|z|^{\rho}), \quad |z| \geq r_0 > 0, \qquad (8.25)$$

holds, with ρ, σ like in (8.23) and (8.24), for $|z| \geq r_0(\varepsilon)$, and $r_0(\varepsilon)$ being sufficiently large.

Proof. According to the Cauchy–Hadamard formula, the radius of convergence of the series (8.22) is $R \geq 0$, where

$$\frac{1}{R} = \limsup_{k \to \infty} (|c_k|)^{\frac{1}{k}} \quad \text{with} \quad c_k = \prod_{i=1}^{m} \left[\frac{(\gamma_i)_k}{k!} \frac{1}{\Gamma(\alpha_i k + \beta_i)} \right]. \tag{8.26}$$

Applying Stirling's asymptotic formulae for large z for the Γ-function (see Remark 6.5), and taking into account (8.26), we have

$$\frac{1}{R} = \limsup_{k \to \infty} (|c_k|)^{\frac{1}{k}} = \lim_{k \to \infty} \prod_{i=1}^{m} \left| e^{\alpha_i} e^{-\alpha_i \ln(\alpha_i k)} \right|, \tag{8.27}$$

thus $1/R = 0$, or $R = \infty$.

The Stirling formula in the form:

$$\ln \Gamma(z) = \left(z - \frac{1}{2} \right) \ln z + \frac{1}{2} \ln(2\pi) + O\left(\frac{1}{z} \right)$$

and the well-known formula for the order of an entire function $\sum_{k=0}^{\infty} c_k z^k$, $\rho = \limsup_{k \to \infty} \frac{k \ln k}{\ln \left(\frac{1}{|c_k|} \right)}$, give now

$$\frac{1}{\rho} = \lim_{k \to \infty} \frac{\ln \left(\frac{1}{|c_k|} \right)}{k \ln k}$$

$$= \lim_{k \to \infty} \left[\sum_{i=1}^{m} \Re(\alpha_i k) \ln k \right] \frac{1 + o(k \ln k)}{k \ln k} = \sum_{i=1}^{m} \Re(\alpha_i),$$

that confirms the validity of the equality (8.23).

Furthermore, the type of the function $\sum_{k=0}^{\infty} c_k z^k$ of order ρ is determined from the equality $(\sigma e \rho)^{\frac{1}{\rho}} = \limsup_{k \to \infty} \left[k^{\frac{1}{\rho}} |c_k|^{\frac{1}{k}} \right]$. Using (8.26) and (8.27) and doing calculations like above, we have

$$\limsup_{k \to \infty} \left[k^{\frac{1}{\rho}} |c_k|^{\frac{1}{k}} \right] = \lim_{k \to \infty} \left[k^{\frac{1}{\rho}} \prod_{i=1}^{m} \left| e^{\alpha_i} e^{-\alpha_i \ln(\alpha_i k)} \right| \right] = \prod_{i=1}^{m} \frac{e^{\Re(\alpha_i)}}{|(\alpha_i)^{\alpha_i}|},$$

from which the following equalities can be deduced:

$$(\sigma e \rho)^{\frac{1}{\rho}} = \prod_{i=1}^{m} \frac{e^{\Re(\alpha_i)}}{|(\alpha_i)^{\alpha_i}|}, \quad (\sigma)^{-\frac{1}{\rho}} = \rho^{\frac{1}{\rho}} \prod_{i=1}^{m} |(\alpha_i)^{\alpha_i}| = \prod_{i=1}^{m} |(\rho\alpha_i)^{\alpha_i}|,$$

that verifies the equality (8.24).

The asymptotic estimate (8.25) follows from the definitions of order and type of an entire function.   □

Alternatively, when any of the parameters $\gamma_1, \ldots, \gamma_m$ is a nonpositive integer, we have the following Theorem.

Theorem 8.5. *If at least one of the parameters $\gamma_1, \ldots, \gamma_m$ is a nonpositive integer, then the multi-index Mittag-Leffler function (8.22) reduces to the finite sum, as follows:*

$$E_{(\alpha_i),(\beta_i)}^{(\gamma_i),m}(z) = \sum_{k=0}^{M} \frac{(\gamma_1)_k \ldots (\gamma_m)_k}{\Gamma(\alpha_1 k + \beta_1) \ldots \Gamma(\alpha_m k + \beta_m)} \frac{z^k}{(k!)^m}. \quad (8.28)$$

Proof. Let $\gamma_{i_1}, \ldots, \gamma_{i_I}$ be the values of γ_i which are nonpositive integers and $M = \min\{-\gamma_{i_1}, \ldots, -\gamma_{i_I}\}$. Then $(-M)_k = 0$ for each $k > M$ and therefore, all the coefficients

$$c_k = \prod_{i=1}^{m} \left[\frac{(\gamma_i)_k}{k!} \frac{1}{\Gamma(\alpha_i k + \beta_i)} \right]$$

are equal to zero when $k > M$. This proves the identity (8.28).   □

Additionally, depending on the other parameters α_i and β_i some of the rest coefficients may be equal to zero, i.e. there exist numbers p and s, such that (8.22) reduces to

$$E_{(\alpha_i),(\beta_i)}^{(\gamma_i),m}(z) = \frac{(\gamma_1)_p \ldots (\gamma_m)_p}{\Gamma(\alpha_1 p + \beta_1) \ldots \Gamma(\alpha_m p + \beta_m)} \frac{z^p}{(p!)^m}$$

$$+ \sum_{k=p+s}^{\infty} \frac{(\gamma_1)_k \ldots (\gamma_m)_k}{\Gamma(\alpha_1 k + \beta_1) \ldots \Gamma(\alpha_m k + \beta_m)} \frac{z^k}{(k!)^m},$$

respectively

$$E_{(\alpha_i),(\beta_i)}^{(\gamma_i),m}(z) = \frac{(\gamma_1)_p \dots (\gamma_m)_p}{\Gamma(\alpha_1 p + \beta_1) \dots \Gamma(\alpha_m p + \beta_m)} \frac{z^p}{(p!)^m}$$

$$+ \sum_{k=p+s}^{M} \frac{(\gamma_1)_k \dots (\gamma_m)_k}{\Gamma(\alpha_1 k + \beta_1) \dots \Gamma(\alpha_m k + \beta_m)} \frac{z^k}{(k!)^m}.$$

This problem has been previously studied in detail by the author for various particular cases of the functions considered above. Here, we give some results in this direction related to the multi-index $(2m)$ Mittag-Leffler functions (8.1), including asymptotic formulae for 'large' values of indices of these functions, obtained in the work [Paneva-Konovska (2012c)].

8.5. Fox's and Wright's Functions

Most of the special functions of Mathematical Physics are particular cases of the generalized hypergeometric functions $_pF_q$, and thus, of the more general Meijer's G-functions [Erdélyi *et al.* (1953, Vol. 1, Chapter 5)]. However, the Mittag-Leffler functions provide an example of special functions which could not be included in the scheme of the Meijer's G-functions, being a more general Fox's H-function. Namely,

$$E_{\alpha,\beta}(z) = H_{1,2}^{1,1}\left[-z \left|\begin{matrix} (0, 1) \\ (0, 1), (1 - \beta, \alpha) \end{matrix}\right.\right]$$

$$= \frac{1}{2\pi i} \int_{\mathcal{L}'} \frac{\Gamma(s)\Gamma(1 - s)}{\Gamma(\beta - \alpha s)}(-z)^{-s} ds \qquad (8.29)$$

with a suitable contour of integration $\mathcal{L}'$ (see e.g. [Kiryakova (1994, Appendix)]), and only for rational $\alpha = p/q$, (8.29) reduces to a G-function. The more general $_p\Psi_q$ functions of Wright also serve as such an example.

Definition 8.3. Under the *Fox's H-function*, we mean a generalized hypergeometric function defined by means of the *Mellin–Barnes type* *contour integral*

$$H_{p,q}^{m,n}(\sigma) = H_{p,q}^{m,n}\left[\sigma \begin{array}{|c}(a_1,\ A_1)\ldots(a_p,\ A_p)\\(b_1,\ B_1)\ldots(b_q,\ B_q)\end{array}\right]$$

$$= H_{p,q}^{m,n}\left[\sigma \begin{array}{|c}(a_k,\ A_k)_1^p\\(b_k,\ B_k)_1^q\end{array}\right] = \frac{1}{2\pi i}\int_{\mathcal{L'}}\mathcal{H}_{p,q}^{m,n}(s)\sigma^s ds,$$

$$(8.30)$$

with an integrand of the form:

$$\mathcal{H}_{p,q}^{m,n}(s) = \frac{\prod_{i=1}^{m}\Gamma(b_i - sB_i)\prod_{j=1}^{n}\Gamma(1 - a_j + sA_j)}{\prod_{i=m+1}^{q}\Gamma(1 - b_i + sB_i)\prod_{j=n+1}^{p}\Gamma(a_j - sA_j)}.$$

The curve $\mathcal{L'}$ is a suitable contour in $\mathbb{C}$, m, n, p and q are integers with $0 \leq m \leq q$ and $0 \leq n \leq p$, the parameters $a_j, b_i \in \mathbb{C}$, A_j, $B_i > 0$, $j = 1,\ldots,p$ and $i = 1,\ldots,q$ and $A_j(b_i + l) \neq B_i(a_j - l' - 1)$, $l, l' = 0, 1, 2, \ldots$. For the various types of contours and conditions for existence and analyticity of the function (8.30) in disks $\subset \mathbb{C}$ with radii $\rho = \prod_{j=1}^{p} A_j^{-A_j} \prod_{i=1}^{q} B_i^{B_i}$, one can see the works [Prudnikov *et al.* (1990); Kiryakova (1994, Appendix); Kilbas *et al.* (2006); Mathai and Haubold (2008)], and so on.

For $A_1 = \cdots = A_p = 1$ and $B_1 = \cdots = B_q = 1$, (8.30) turns into the more popular Meijer's G-function, see [Erdélyi *et al.* (1953, Vol. 1, Chapter 5)]. The G- and H-functions encompass almost all elementary and special functions and this makes the knowledge on them very useful. Observe that the generalized hypergeometric functions $_pF_q$ are special cases of the G-function:

$$_pF_q(a_1,\ldots,a_p; b_1,\ldots,b_q;\sigma)$$

$$= \sum_{k=0}^{\infty}\frac{(a_1)_k\ldots(a_p)_k}{(b_1)_k\ldots(b_q)_k}\frac{\sigma^k}{k!}$$

$$= \frac{\prod_{i=1}^{q}\Gamma(b_i)}{\prod_{i=1}^{p}\Gamma(a_i)}G_{p,q+1}^{1,p}\left[-\sigma \begin{array}{|c}1 - a_1,\ldots,1 - a_p\\0, 1 - b_1,\ldots,1 - b_q\end{array}\right], \quad (8.31)$$

while the Mittag-Leffler functions (7.2) with irrational parameters $\alpha > 0$ and the Wright generalized hypergeometric functions $_p\Psi_q$ with irrational $A_j, B_i > 0$ give examples of H-functions, not reducible to G-functions, see e.g. representation (8.29), and

$$
{}_p\Psi_q \left[\begin{matrix} (a_1,\ A_1) \dots (a_p,\ A_p) \\ (b_1,\ B_1) \dots (b_q,\ B_q) \end{matrix} \middle| \sigma \right]
$$

$$
= \sum_{k=0}^{\infty} \frac{\Gamma(a_1 + kA_1) \dots \Gamma(a_p + kA_p)}{\Gamma(b_1 + kB_1) \dots \Gamma(b_q + kB_q)} \frac{\sigma^k}{k!}
$$

$$
= H_{p,\ q+1}^{1,\ p} \left[-\sigma \middle| \begin{matrix} (1 - a_1,\ A_1), \dots, (1 - a_p,\ A_p) \\ (0,1), (1 - b_1,\ B_1), \dots, (1 - b_q,\ B_q) \end{matrix} \right]. \quad (8.32)
$$

However, for $A_1 = \cdots = A_p = 1$ and $B_1 = \cdots = B_q = 1$, *cf.* (8.31),

$$
{}_p\Psi_q \left[\begin{matrix} (a_1,\ 1) \dots (a_p,\ 1) \\ (b_1,\ 1) \dots (b_q,\ 1) \end{matrix} \middle| \sigma \right]
$$

$$
= \frac{\prod_{i=1}^{p} \Gamma(a_i)}{\prod_{i=1}^{q} \Gamma(b_i)} {}_pF_q(a_1, \dots, a_p; b_1, \dots, b_q; \sigma)
$$

$$
= G_{p,\ q+1}^{1,\ p} \left[-\sigma \middle| \begin{matrix} 1 - a_1, \dots, 1 - a_p \\ 0, 1 - b_1, \dots, 1 - b_q \end{matrix} \right]. \quad (8.33)
$$

8.6. The $3m$-parametric Multi-index Mittag-Leffler Functions as Fox's and Wright's Functions

Furthermore, let us emphasize the place of the introduced functions (8.22) among the known special functions, especially in the class of the Wright generalized hypergeometric functions and Fox H-functions (see [e.g. Prudnikov *et al.* (1990); Kiryakova (1994, Appendix)]).

For this purpose, letting $\gamma_i \in \mathbb{C}$ and $\Re(\gamma_i) > 0$, $i = 1, \dots, m$, we firstly define two numerical sets as it is given below:

$$
S_l = \{s : s = -k \quad (k \in \mathbb{N}_0)\},
$$

$$
S_r = \{s : s = l + \gamma_i \quad (l \in \mathbb{N}_0)\}.
$$

Remark 8.6. The intersection of the sets S_l and S_r is empty, i.e. $S_l \cap S_r = \varnothing$. Moreover, if

$$\tilde{\gamma} = \min_{i=1/m} \Re(\gamma_i), \tag{8.34}$$

then the set S_l lies on the left-hand side of the strip

$$S = \{s : s \in \mathbb{C}, \ 0 < \gamma' \leq \Re(s) \leq \gamma'' < \tilde{\gamma}\}, \tag{8.35}$$

while the set S_r lies on its right.

Here, we prove the following new representations.

Theorem 8.6. *Let $\alpha_i > 0$, $\beta_i, \gamma_i \in \mathbb{C}$ and $\Re(\gamma_i) > 0$ for $i = 1, \ldots, m$. Then the multi-index Mittag-Leffler functions* (8.22) *are expressed by Wright's generalized hypergeometric functions as well as by Fox's H-functions in the form:*

$$E_{(\alpha_i),(\beta_i)}^{(\gamma_i),m}(z)$$

$$= A_m\Psi_{2m-1}\left[\begin{matrix} (\gamma_1, 1), \ldots, (\gamma_m, 1) \\ (\beta_1, \alpha_1), \ldots, (\beta_m, \alpha_m), (1,1), \ldots, (1,1) \end{matrix} \middle| z \right]$$

$$= A\, H_{m,2m}^{1,m}\left[-z \middle| \begin{matrix} (1-\gamma_1, 1), \ldots, (1-\gamma_m, 1) \\ [(0,1), (1-\beta_i, \alpha_i)]_1^m \end{matrix} \right], \tag{8.36}$$

with $A = [\prod_{i=1}^m \Gamma(\gamma_i)]^{-1}$. They have the following Mellin–Barnes type contour integral representation, extending the integral formula (8.29):

$$E_{(\alpha_i),(\beta_i)}^{(\gamma_i),m}(z) = \frac{A}{2\pi i} \int_{\mathcal{L}'} \mathcal{H}_{m,2m}^{1,m}(s)(-z)^s ds$$

$$= \frac{A}{2\pi i} \int_{\mathcal{L}} \mathcal{H}_{m,2m}^{1,m}(-s)(-z)^{-s} ds, \tag{8.37}$$

for $|\arg(-z)| < \pi$, where

$$\mathcal{H}_{m,2m}^{1,m}(s) = \frac{\Gamma(-s)\prod_{i=1}^m \Gamma(\gamma_i + s)}{[\Gamma(1+s)]^{m-1}\prod_{i=1}^m \Gamma(\beta_i + s\alpha_i)} \tag{8.38}$$

and $\mathcal{L}$ is an arbitrary contour in $\mathbb{C}$ running from $-i\infty$ to $+i\infty$ in a way that the poles $s = -k$ ($k \in \mathbb{N}_0$) of $\Gamma(s)$ lie to the left of $\mathcal{L}$ and *the poles $s = l + \gamma_i$ $(l \in \mathbb{N}_0)$ of $\Gamma(\gamma_i - s)$ $(i = 1, \ldots, m)$ to the right of it.*

Proof. According to Remark 8.6, none of the poles of $\Gamma(s)$ and $\Gamma(\gamma_i - s)$ are in the strip S, given by (8.35). Moreover, the poles $s = -k$ $(k \in \mathbb{N}_0)$ of $\Gamma(s)$ lie to the left of this strip and the poles $s = l + \gamma_i$ $(l \in \mathbb{N}_0)$ of $\Gamma(\gamma_i - s)$ $(i = 1, \ldots, m)$ to its right.

Let us consider the right-hand side of (8.37):

$$I(z) = \frac{A}{2\pi i} \int_{\mathcal{L}} \mathcal{H}^{1,m}_{m,2m}(-s)(-z)^{-s} ds. \tag{8.39}$$

Calculating the residues of the integrand of (8.39) at the simple poles $s_k = -k$, $k = 0, 1, 2, \ldots$, and taking into account the asymptotic formula (see e.g. [Erdélyi *et al.* (1953, Vol. 1, 1.1 (8))]):

$$\Gamma(s) = \frac{(-1)^k}{k!(s+k)}[1 + O(s+k)] \quad (s \to -k;\ k = 0, 1, 2, \ldots),$$

we have

$$
\begin{aligned}
I(z) &= A \sum_{k=0}^{\infty} \mathrm{Res}_{s=-k} \left\{ \mathcal{H}^{1,m}_{m,2m}(-s)(-z)^{-s} \right\} \\
&= A \sum_{k=0}^{\infty} \mathrm{Res}_{s=-k} \left\{ \frac{\Gamma(s) \prod_{i=1}^{m} \Gamma(\gamma_i - s)}{[\Gamma(1-s)]^{m-1} \prod_{i=1}^{m} \Gamma(\beta_i - s\alpha_i)} (-z)^{-s} \right\} \\
&= A \sum_{k=0}^{\infty} \left\{ \frac{(-1)^k \prod_{i=1}^{m} \Gamma(\gamma_i + k)}{k![\Gamma(1+k)]^{m-1} \prod_{i=1}^{m} \Gamma(\beta_i + k\alpha_i)} (-z)^k \right\} \\
&= A \sum_{k=0}^{\infty} \frac{\prod_{i=1}^{m} \Gamma(\gamma_i + k)}{\prod_{i=1}^{m} \Gamma(\beta_i + k\alpha_i)} \frac{z^k}{(k!)^m} = E^{(\gamma_i),m}_{(\alpha_i),(\beta_i)}(z), \tag{8.40}
\end{aligned}
$$

that proves (8.37).

Furthermore, writing (8.22) in the form:

$$E^{(\gamma_i),m}_{(\alpha_i),(\beta_i)}(z) = A \sum_{k=0}^{\infty} \frac{\Gamma(k+\gamma_1)\ldots\Gamma(k+\gamma_m)}{\Gamma(\alpha_1 k + \beta_1)\cdots\Gamma(\alpha_m k + \beta_m)(\Gamma(k+1))^{m-1}} \frac{z^k}{k!}$$

and comparing with (8.32), we obtain (8.36) that completes the proof of the theorem.   □

The representation (8.36) of the multi-index Mittag-Leffler functions as Fox's H-functions allows to describe their asymptotic behavior when $z \to 0$, as well as $z \to \infty$. In the case of Mittag-Leffler function (7.2) ($m = 1$), Dzrbashjan (1966) established different asymptotic formulas for $|z| \to \infty$, valid in different parts of the complex plane and under different conditions on α and β. For example, if $1/\alpha := \rho > 1/2$ and inside angle domains, (7.2) is $\approx \rho z^{\rho(1-\mu)} \exp(z^\rho)$. An asymptotic estimate in the case $m > 1$ is given by (8.25) and in more detailed situations, the asymptotics of the multi-index Mittag-Leffler functions could be found from their interpretation as ${}_p\Psi_q$ functions.

The Mellin transform of a function $f(t)$ of a real variable $t \in \mathbb{R}^+ = (0, \infty)$ is defined by

$$(\mathcal{M}f)(s) = \mathcal{M}[f(t)](s) = F(s) = \int_0^\infty f(t)t^{s-1}dt \quad (s \in S \subset \mathbb{C}),$$

(8.41)

S being a suitable vertical strip (see, for example, the book of Titchmarsh (1986)), and the inverse Mellin transform is given for $t \in \mathbb{R}^+$ by the formula ($\gamma = \Re(s)$):

$$\left(\mathcal{M}^{-1}F\right)(t) = \mathcal{M}^{-1}[F(s)](t) = \frac{1}{2\pi i}\int_{\gamma - i\infty}^{\gamma + i\infty} F(s)t^{-s}ds, \quad 0 < t < \infty.$$

(8.42)

Corollary 8.1. *Let $\alpha_i > 0$, $\beta_i, \gamma_i \in \mathbb{C}$ and $\Re(\gamma_i) > 0$ for $i = 1, \ldots, m$. Then the Mellin transform of the $3m$-multi-index Mittag-Leffler function has the form:*

$$\mathcal{M}\left[E_{(\alpha_i),(\beta_i)}^{(\gamma_i),m}(-t)\right](s) = A\,\mathcal{H}_{m,2m}^{1,m}(-s),$$

(8.43)

with $t > 0$, $0 < \Re(s) < \tilde{\gamma}$, $\tilde{\gamma}$ and $\mathcal{H}_{m,2m}^{1,m}$ like in (8.34), resp. (8.38), and $A = [\prod_{i=1}^m \Gamma(\gamma_i)]^{-1}$.

Proof. In particular, if $\mathcal{L}$ is the straight line $\Re(s) = \gamma$, $0 < \gamma < \tilde{\gamma}$, and taking $-z = t \in (0, \infty)$, then the relation (8.37) leads to

$$E^{(\gamma_i),m}_{(\alpha_i),(\beta_i)}(-t) = \frac{A}{2\pi i} \int_{\gamma-i\infty}^{\gamma+i\infty} \mathcal{H}^{1,m}_{m,2m}(-s)(t)^{-s} ds, \tag{8.44}$$

for $0 < t < \infty$. The relation (8.44), obtained above, means that the function $E^{(\gamma_i),m}_{(\alpha_i),(\beta_i)}(-t)$ is the inverse Mellin transform of the function $A\,\mathcal{H}^{1,m}_{m,2m}(-s)$. Therefore, the direct Mellin transform of this function is given by (8.43), that should be proved. $\qquad\square$

8.7. Special Cases of the Multi-index Mittag-Leffler Functions

Let us mention some interesting special cases of the multi-index Mittag-Leffler functions, with $2m$ and $3m$ indices, respectively.

Case 1: If $m = 1$, the formula (8.22) gives the three-parametric Mittag-Leffler type function, also known as Prabhakar function (7.8):

$$E^\gamma_{\alpha,\beta}(z) = E^{(\gamma),1}_{(\alpha),(\beta)}(z) = \frac{1}{\Gamma(\gamma)} \, {}_1\Psi_1 \left[\begin{matrix} (\gamma, 1) \\ (\beta, \alpha) \end{matrix} \middle| z \right].$$

In addition, when $\gamma = 1$, we get the classical Mittag-Leffler function and its special cases:

$$E_{\alpha,\beta}(z) = E^{(1),1}_{(\alpha),(\beta)}(z) = {}_1\Psi_1 \left[\begin{matrix} (1, 1) \\ (\beta, \alpha) \end{matrix} \middle| z \right],$$

$$E_\alpha(z) = E^{(1),1}_{(\alpha),(1)}(z) = {}_1\Psi_1 \left[\begin{matrix} (1, 1) \\ (1, \alpha) \end{matrix} \middle| z \right],$$

$$E_1(z) = \exp(z) = E^{(1),1}_{(1),(1)}(z) = {}_1\Psi_1 \left[\begin{matrix} (1, 1) \\ (1, 1) \end{matrix} \middle| z \right].$$

Case 2: If $\gamma_1 = \cdots = \gamma_m = 1$, the definition (8.22) gives the $(2m\text{-})$ multi-index Mittag-Leffler functions (8.1), introduced by Yakubovich and Luchko [1994], Luchko [1999] and studied in a series of Kiryakova's papers. More precisely, putting $1/\rho_i$ in (8.22) instead of α_i, we obtain:

$$E_{\left(\frac{1}{\rho_i}\right),(\beta_i)}(z) = E^{(1),m}_{\left(\frac{1}{\rho_i}\right),(\beta_i)}(z) = E^{(1,\ldots,1),m}_{\left(\frac{1}{\rho_i}\right),(\beta_i)}(z)$$

$$= {}_m\Psi_{2m-1}\left[\begin{array}{c}(1,\ 1),\ldots,(1,\ 1)\\ (\beta_i,\ \alpha_i)_1^m,(1,1),\ldots,(1,1)\end{array}\middle|\ z\right]$$

$$= {}_1\Psi_m\left[\begin{array}{c}(1,\ 1)\\ (\beta_i,\ \alpha_i)_1^m\end{array}\middle|\ z\right].\tag{8.45}$$

Then, naturally, the formulae (8.23) and (8.24) reduce to (8.2), obtained by Kiryakova [2000].

As it is well known, the last function is connected with the *hyper-Bessel function*, defined by the formulae (6.5)–(6.7). More precisely, taking into consideration the formula:

$$j^{(m)}_{\nu_1,\ldots,\nu_m}(z) = {}_0F_m\left((\nu_k+1)\big|_1^m; -\left(\frac{z}{m+1}\right)^{m+1}\right)$$

as well, the following relation holds (for details see [Kiryakova (1994, Appendix (D. 3), Chapter 3)]):

$$J^{(m-1)}_{\nu_1,\ldots,\nu_{m-1}}(z) = \frac{\left(\frac{z}{m}\right)^{\nu_1+\cdots+\nu_{m-1}}}{\prod_{i=1}^{m-1}\Gamma(\nu_i+1)}\ {}_0F_m\left((\nu_k+1)\big|_1^{m-1}; -\left(\frac{z}{m}\right)^m\right)$$

$$= \left(\frac{z}{m}\right)^{\sum_{i=1}^{m-1}\nu_i}E^m_{(1,\ldots,1),(\nu_1+1,\ldots,\nu_{m-1}+1,1)}\left(-\left(\frac{z}{m}\right)^m\right)$$

$$= \left(\frac{z}{m}\right)^{\sum_{i=1}^{m-1}\nu_i}E^{(1,\ldots,1),m}_{(1,\ldots,1),(\nu_1+1,\ldots,\nu_{m-1}+1,1)}\left(-\left(\frac{z}{m}\right)^m\right).$$

$$\tag{8.46}$$

Case 3: Again, let $\gamma_1 = \cdots = \gamma_m = 1$. A special case (for $m = 2$) is the generalized Wright's functions with 2×2 indices (see, for their applications as scale-invariant solutions of fractional order PDEs, for example, in Luchko and Gorenflo [1999]) and Dzrbashjan's Mittag-Leffler type functions from the book [Dzrbashjan (1966)], denoted as

$E_{(1/\rho_1,1/\rho_2),(\mu_1,\mu_2)}$, i.e.

$$E_{\left(\frac{1}{\rho_1},\frac{1}{\rho_2}\right),(\mu_1,\mu_2)}(z) = E^{(1,1),\,2}_{\left(\frac{1}{\rho_1},\frac{1}{\rho_2}\right),(\mu_1,\mu_2)}(z) = {}_1\Psi_2 \left[\begin{matrix} (1,\,1) \\ \left(\mu_i,\,\frac{1}{\rho_i}\right)_1^2 \end{matrix} \middle| z \right].$$

(8.47)

Further, for $m \geq 2$, the function (8.45), as well as (8.22), is the generalized Lommel–Wright function with four indices ($\mu > 0$, $q \in \mathbb{N}$, $\nu, \lambda \in \mathbb{C}$):

$$J^{\mu,q}_{\nu,\lambda}(z) = \left(\frac{z}{2}\right)^{\nu+2\lambda} E^{(1,1,\dots,1),\,q+1}_{(\mu,1,\dots,1),(\nu+\lambda+1,\lambda+1,\dots,\lambda+1)} \left(-\left(\frac{z}{2}\right)^2 \right)$$

$$= \left(\frac{z}{2}\right)^{\nu+2\lambda} {}_1\Psi_{q+1} \left[\begin{matrix} (1,\,1) \\ (\lambda+1,1)^q_1,(\lambda+\nu+1,\,\mu) \end{matrix} \middle| -\left(\frac{z}{2}\right)^2 \right].$$

(8.48)

This is an example of a multi-index Mittag-Leffler function with arbitrary $m = q+1$. Some particular cases of (8.48) are listed below.

Obviously, for $q = 1$, the special function (8.48) turns into the 3-index generalization (6.3) of the Bessel function $J_\nu(z)$:

$$J^{\mu}_{\nu,\lambda}(z) = \left(\frac{z}{2}\right)^{\nu+2\lambda} E^{(1,1),\,2}_{(\mu,1),(\nu+\lambda+1,\lambda+1)} \left(-\left(\frac{z}{2}\right)^2 \right)$$

$$= \left(\frac{z}{2}\right)^{\nu+2\lambda} {}_1\Psi_2 \left[\begin{matrix} (1,\,1) \\ (\lambda+1,1),(\lambda+\nu+1,\,\mu) \end{matrix} \middle| -\left(\frac{z}{2}\right)^2 \right].$$

(8.49)

For particular choices of the other parameters λ and μ, we obtain results for more special cases as follows.

Let $\lambda = 0$, then the special function (8.48) gives the generalization of the Bessel–Clifford function $C_\nu(z) = z^{-\nu/2} J_\nu(2\sqrt{z})$, introduced by Edward Maitland Wright with (6.2):

$$J^{\mu}_{\nu}(z) = E^{(1,1),\,2}_{(\mu,1),(\nu+1,1)}(-z) = {}_0\Psi_1 \left[\begin{matrix} -- \\ (\nu+1,\,\mu) \end{matrix} \middle| -z \right].$$

(8.50)

Additionally, if $\mu = 1$, then from (8.50) we get the classical Bessel function:

$$J_\nu(z) = J_\nu^{(1)}(z) = \left(\frac{z}{2}\right)^\nu E_{(1,1),(\nu+1,1)}^{(1,1),\,2}\left(-\left(\frac{z}{2}\right)^2\right)$$

$$= \left(\frac{z}{2}\right)^\nu {}_0\Psi_1\left[\begin{array}{c} -\,- \\ (\nu+1,\,1) \end{array}\Big| -\left(\frac{z}{2}\right)^2\right]. \tag{8.51}$$

8.8. Fractional Riemann–Liouville Integral and Derivative

The notion FC or fractional analysis is used for the extension of the calculus (analysis), when the order of integration and differentiation can be an arbitrary number (fractional, irrational or complex), that is, not obligatorily integer. For its theory and applications, see the *FC encyclopedia* [Samko *et al.* (1993)]. The most popular definition for the integration of order $\lambda \in \mathbb{C}$ $(\Re(\lambda) > 0)$, is the *Riemann–Liouville (RL) fractional integral*

$$R^\lambda f(z) = \frac{1}{\Gamma(\lambda)} \int_0^z (z-t)^{\lambda-1} f(t) dt = \frac{z^\lambda}{\Gamma(\lambda)} \int_0^1 (1-\tau)^{\lambda-1} f(z\tau) d\tau. \tag{8.52}$$

Then, the *RL fractional derivative* of order $\lambda \in \mathbb{C}$ $(\Re(\lambda) > 0)$, is defined as a composition of a derivative of integer order and an integral of fractional order of the form (8.52), namely:

$$D^\lambda f(z) := D^n R^{n-\lambda} f(z), \tag{8.53}$$

where $n := [\Re(\lambda)] + 1 > \Re(\lambda)$, $[\Re(\lambda)] =$ integer part of $\Re(\lambda)$.

In this section, we consider RL fractional integrals and derivatives (8.52) and (8.53) of the multi-index Mittag-Leffler function (8.22). To this purpose, we first need the derivatives of integer order. The differentiation of the multi-index Mittag-Leffler function (8.22) is given by the following elementary assertion.

Theorem 8.7. *Let* $\alpha_i, \beta_i, \gamma_i, \omega \in \mathbb{C}, \Re(\alpha_i) > 0$ *and* $\Re(\beta_{i_0}) > 0, i = 1, \ldots, m, 1 \le i_0 \le m$ *and* $i_0 \in \mathbb{N}$, *then for any* $n \in \mathbb{N}$ *the following* *identity holds:*

$$D^n \left[z^{\beta_{i_0}-1} E^{(\gamma_i),m}_{(\alpha_i),(\beta_i)} (\omega z^{\alpha_{i_0}}) \right]$$

$$= \left(\frac{d}{dz} \right)^n \left[z^{\beta_{i_0}-1} E^{(\gamma_i),m}_{(\alpha_i),(\beta_i)} (\omega z^{\alpha_{i_0}}) \right]$$

$$= z^{\beta_{i_0}-n-1} E^{(\gamma_i),m}_{(\alpha_i),(\widetilde{\beta}_i)} (\omega z^{\alpha_{i_0}}), \quad |\arg z| < \pi, \qquad (8.54)$$

with $\widetilde{\beta}_{i_0} = \beta_{i_0} - n$ *and* $\widetilde{\beta}_i = \beta_i$, *if* $i \neq i_0$.

Proof. Using (8.22) and taking term-by-term differentiation under the summation sign (which is possible in accordance with the uniform convergence of the series (8.22) in any compact subset of $\mathbb{C}$ and the differentiability of the power function for $|\arg z| < \pi$), we have

$$\left(\frac{d}{dz} \right)^n \left[z^{\beta_{i_0}-1} E^{(\gamma_i),m}_{(\alpha_i),(\beta_i)} (\omega z^{\alpha_{i_0}}) \right]$$

$$= \sum_{k=0}^{\infty} \frac{(\gamma_1)_k \cdots (\gamma_m)_k}{\Gamma(\alpha_1 k + \beta_1) \ldots \Gamma(\alpha_m k + \beta_m)} \frac{\omega^k \left(\frac{d}{dz} \right)^n \left[z^{\alpha_{i_0} k + \beta_{i_0}-1} \right]}{(k!)^m},$$

from which, applying the relation

$$\left(\frac{d}{dz} \right)^n \left[z^{\alpha_{i_0} k + \beta_{i_0}-1} \right] = \frac{\Gamma(\alpha_{i_0} k + \beta_{i_0})}{\Gamma(\alpha_{i_0} k + \beta_{i_0} - n)} z^{\alpha_{i_0} k + \beta_{i_0}-n-1},$$

the truth of the identity (8.54) follows immediately. $\square$

Remark 8.7. Additionally, if $\alpha_{i_0}, \beta_{i_0} \in \mathbb{R}$, then (8.54) is correct not only for $|\arg z| < \pi$ but also for all the complex values of z, probably except for the origin.

Remark 8.8. In particular, as a result of (8.54), we obtain the following identities:

$$D^n \left[z^{\beta-1} E^{\gamma}_{\alpha,\beta}(\omega z^{\alpha}) \right] = z^{\beta-n-1} E^{\gamma}_{\alpha,\beta-n}(\omega z^{\alpha}),$$

$$D^n \left[z^{\beta_{i_0}-1} E^m_{(\alpha_i),(\beta_i)} (\omega z^{\alpha_{i_0}}) \right] = z^{\beta_{i_0}-n-1} E^m_{(\alpha_i),(\widetilde{\beta}_i)} (\omega z^{\alpha_{i_0}}),$$

with $\widetilde{\beta}_{i_0} = \beta_{i_0} - n$ and $\widetilde{\beta}_i = \beta_i$, if $i \neq i_0$. These results follow, taking into account (7.8) and (8.1), if $m = 1$, respectively,

$\gamma_1 = \cdots = \gamma_m = 1$. The above-mentioned relations are well known, for details see the works of [Kilbas *et al.* (2004); Kiryakova (2000)].

Corollary 8.2. *Let* $\alpha_i, \beta_i, \gamma_i, \omega \in \mathbb{C}, \Re(\alpha_i) > 0$ *and* $\Re(\beta_{i_0}) > 0$, $i = 1, \ldots, m, \ 1 \le i_0 \le m$ *and* $i_0 \in \mathbb{N}$, *then*

$$\int_0^z t^{\beta_{i_0}-1} E^{(\gamma_i),m}_{(\alpha_i),(\beta_i)} \left(\omega t^{\alpha_{i_0}}\right) dt = z^{\beta_{i_0}} E^{(\gamma_i),m}_{(\alpha_i),(\widetilde{\beta}_i)} \left(\omega z^{\alpha_{i_0}}\right), \qquad (8.55)$$

with $\widetilde{\beta}_i = \beta_i$, *if* $i \neq i_0$, *and* $\widetilde{\beta}_{i_0} = \beta_{i_0} + 1$.

Proof. Setting $n = 1$ and β_{i_0} instead of $\beta_{i_0} - 1$, the relation (8.54) leads to the equality

$$D \left[z^{\beta_{i_0}} E^{(\gamma_i),m}_{(\alpha_i),(\widetilde{\beta}_i)} \left(\omega z^{\alpha_{i_0}}\right) \right] = z^{\beta_{i_0}-1} E^{(\gamma_i),m}_{(\alpha_i),(\beta_i)} \left(\omega z^{\alpha_{i_0}}\right).$$

The last formula shows that the function $z^{\beta_{i_0}} E^{(\gamma_i),m}_{(\alpha_i),(\widetilde{\beta}_i)} \left(\omega z^{\alpha_{i_0}}\right)$ is a primitive function of $z^{\beta_{i_0}-1} E^{(\gamma_i),m}_{(\alpha_i),(\beta_i)} \left(\omega z^{\alpha_{i_0}}\right)$, that verifies (8.55). $\quad\square$

Theorem 8.8. *Let* $\alpha_i, \beta_i, \gamma_i, \omega, \lambda \in \mathbb{C}$ *and* $\Re(\alpha_i), \Re(\beta_{i_0}), \Re(\lambda) > 0$, $i = 1, \ldots, m, \ 1 \le i_0 \le m$ *and* $i_0 \in \mathbb{N}$, *then the following identity holds:*

$$R^\lambda \left[z^{\beta_{i_0}-1} E^{(\gamma_i),m}_{(\alpha_i),(\beta_i)} \left(\omega z^{\alpha_{i_0}}\right) \right] = z^{\beta_{i_0}+\lambda-1} E^{(\gamma_i),m}_{(\alpha_i),(\widetilde{\beta}_i)} \left(\omega z^{\alpha_{i_0}}\right), \qquad (8.56)$$

with $\widetilde{\beta}_{i_0} = \beta_{i_0} + \lambda$ *and* $\widetilde{\beta}_i = \beta_i$, *if* $i \neq i_0$.

Proof. Putting $f(t) = t^{\beta_{i_0}-1} E^{(\gamma_i),m}_{(\alpha_i),(\beta_i)} \left(\omega t^{\alpha_{i_0}}\right)$, we obtain

$$f(z\tau) = z^{\beta_{i_0}-1} \tau^{\beta_{i_0}-1} E^{(\gamma_i),m}_{(\alpha_i),(\beta_i)} \left(\omega z^{\alpha_{i_0}} \tau^{\alpha_{i_0}}\right)$$

$$= z^{\beta_{i_0}-1} \sum_{k=0}^{\infty} \frac{(\gamma_1)_k \cdots (\gamma_m)_k}{\Gamma(\alpha_1 k + \beta_1) \ldots \Gamma(\alpha_m k + \beta_m)} \frac{(\omega z^{\alpha_{i_0}})^k}{(k!)^m} \tau^{\alpha_{i_0} k + \beta_{i_0}-1}.$$

Now, using the formula (8.52) and taking into account the relation:

$$\int_0^1 (1-\tau)^{\lambda-1} \tau^{\alpha_{i_0} k + \beta_{i_0}-1} d\tau = B\left(\lambda, \alpha_{i_0} k + \beta_{i_0}\right) = \frac{\Gamma(\lambda)\Gamma(\alpha_{i_0} k + \beta_{i_0})}{\Gamma(\alpha_{i_0} k + \beta_{i_0} + \lambda)},$$

we complete the proof of the theorem. $\quad\square$

Theorem 8.9. *Let* $\alpha_i, \beta_i, \gamma_i, \omega, \lambda \in \mathbb{C}$ *and* $\Re(\alpha_i), \Re(\beta_{i_0}), \Re(\lambda) >$ 0, $i = 1, \ldots, m$, $1 \le i_0 \le m$, $i_0 \in \mathbb{N}$, *then the following identity holds:*

$$D^{\lambda} \left[z^{\beta_{i_0}-1} E^{(\gamma_i),m}_{(\alpha_i),(\beta_i)} \left(\omega z^{\alpha_{i_0}} \right) \right] = z^{\beta_{i_0}-\lambda-1} E^{(\gamma_i),m}_{(\alpha_i),(\widetilde{\beta}_i)} \left(\omega z^{\alpha_{i_0}} \right), \quad (8.57)$$

with $\widetilde{\beta}_i = \beta_i$, *if* $i \ne i_0$, *and* $\widetilde{\beta}_{i_0} = \beta_{i_0} - \lambda$.

Proof. Using (8.53) and (8.54), we apply consecutively $R^{n-\lambda}$ and D^n subsequently and obtain the following results:

$$R^{n-\lambda} \left[z^{\beta_{i_0}-1} E^{(\gamma_i),m}_{(\alpha_i),(\beta_i)} \left(\omega z^{\alpha_{i_0}} \right) \right] = z^{\beta_{i_0}+n-\lambda-1} E^{(\gamma_i),m}_{(\alpha_i),(\beta'_i)} \left(\omega z^{\alpha_{i_0}} \right),$$

$$D^n \left[z^{\beta_{i_0}+n-\lambda-1} E^{(\gamma_i),m}_{(\alpha_i),(\beta'_i)} \left(\omega z^{\alpha_{i_0}} \right) \right] = z^{\beta_{i_0}-\lambda-1} E^{(\gamma_i),m}_{(\alpha_i),(\widetilde{\beta}_i)} \left(\omega z^{\alpha_{i_0}} \right),$$

with $\widetilde{\beta}_i = \beta'_i = \beta_i$, if $i \ne i_0$, and $\beta'_{i_0} = \beta_{i_0} + n - \lambda$, $\widetilde{\beta}_{i_0} = \beta'_{i_0} - n = \beta_{i_0} - \lambda$, that confirms the correctness of the relation (8.57). □

8.9. Generalized Fractional Erdélyi–Kober Integrals and Derivatives

Along with the classical R–L definitions of fractional order operators, many other modifications and their generalizations have been used in the literature. The most useful operators of the classical FC however seem to be the Erdélyi–Kober operators, for details, see e.g. [Kiryakova (1994)]. The *Erdélyi–Kober fractional integral operator* is defined as

$$\begin{aligned} I^{\beta,\lambda}_{\mu} f(z) &= \left[z^{-(\beta+\lambda)} R^{\lambda} z^{\beta} f\left(z^{\frac{1}{\mu}} \right) \right]_{z \to z^{\mu}} \\ &= \frac{z^{-\mu(\beta+\lambda)}}{\Gamma(\lambda)} \int_0^z (z^{\mu} - t^{\mu})^{\lambda-1} t^{\beta\mu} f(t) d(t^{\mu}) \\ &= \frac{1}{\Gamma(\lambda)} \int_0^1 (1-\tau)^{\lambda-1} \tau^{\beta} f\left(z\tau^{\frac{1}{\mu}} \right) d\tau, \quad (8.58) \end{aligned}$$

and the corresponding *Erdélyi–Kober differential operator* can be written symbolically as

$$D^{\beta,\lambda}_{\mu} f(z) = \left[z^{-\beta} D^{\lambda} z^{\beta+\lambda} f\left(z^{\frac{1}{\mu}} \right) \right]_{z \to z^{\mu}}, \quad (8.59)$$

where λ, $\mu > 0$ and β are real parameters $(D_\mu^{\beta,0} f(z) = I_\mu^{\beta,0} f(z) = f(z)$ by default) and $D_\mu^{\beta,\lambda} I_\mu^{\beta,\lambda} = Id$. The cases $\mu = 2$ are introduced by Sneddon, while those originally considered by Kober and Erdélyi have $\mu = 1$. The Erdélyi–Kober operators (8.58) and (8.59) have numerous applications emphasized yet by Sneddon, and are widely used in FC and fractional order differential equations to model various type of physical, economical and many more processes. See for example [Pagnini (2012)], considering the so-called Erdélyi–Kober fractional diffusion, a family of diffusive processes governed by the generalized grey Brownian motion (ggBm).

The operator (8.58) is essentially used in this section. First, we provide the Erdélyi–Kober fractional integrals of the functions (7.8) and (8.22).

Lemma 8.4. *Let $i = 1, 2, \ldots, m$, i_0 be a fixed natural number, $1 \leq i_0 \leq m$, $\omega \in \mathbb{C}$, $\omega \neq 0$, and α, β, α_i, β_{i_0}, $\lambda > 0$. Then*

$$I_{\frac{1}{\alpha}}^{\beta-1,\lambda} E_{\alpha,\beta}^\gamma(\omega z) = E_{\alpha,\beta+\lambda}^\gamma(\omega z), \tag{8.60}$$

$$I_{\frac{1}{\alpha_{i_0}}}^{\beta_{i_0}-1,\lambda} E_{(\alpha_i),(\beta_i)}^{(\gamma_i),m}(\omega z) = E_{(\alpha_i),(\widetilde{\beta}_i)}^{(\gamma_i),m}(\omega z), \tag{8.61}$$

with

$$\widetilde{\beta}_{i_0} = \beta_{i_0} + \lambda; \quad \widetilde{\beta}_i = \beta_i, \quad i \neq i_0.$$

Proof. To prove (8.60), we set $f(z) = E_{\alpha,\beta}^\gamma(\omega z)$, and integrating term-by-term, (8.58) gives

$$I_{\frac{1}{\alpha}}^{\beta-1,\lambda} f(z) = \frac{1}{\Gamma(\lambda)} \int_0^1 (1-\tau)^{\lambda-1} \tau^{\beta-1} f(z\tau^\alpha) d\tau$$

$$= \frac{1}{\Gamma(\lambda)} \int_0^1 (1-\tau)^{\lambda-1} \tau^{\beta-1} \sum_{k=0}^\infty \frac{(\gamma)_k}{\Gamma(\alpha k + \beta)} \frac{(\omega z)^k \tau^{\alpha k}}{k!} d\tau$$

$$= \frac{1}{\Gamma(\lambda)} \sum_{k=0}^\infty \frac{(\gamma)_k}{\Gamma(\alpha k + \beta)} \frac{(\omega z)^k}{k!} \int_0^1 (1-\tau)^{\lambda-1} \tau^{\alpha k + \beta - 1} d\tau$$

$$= \sum_{k=0}^\infty \frac{(\gamma)_k}{\Gamma(\alpha k + \beta + \lambda)} \frac{(\omega z)^k}{k!} = E_{\alpha,\beta+\lambda}^\gamma(\omega z),$$

that proves (8.60). Analogously, putting $f(z) = E_{(\alpha_i),(\beta_i)}^{(\gamma_i),m}(\omega z)$ and applying (8.58) with $\mu = 1/\alpha_{i_0}$ and $\beta = \beta_{i_0} - 1$, we obtain (8.61).

$\square$

Now, we briefly recall some definitions of the operators of the so-called generalized fractional calculus (GFC) of Kiryakova [1994, 2000].

Definition 8.4. Let $m \geq 1$ be an integer, $\lambda_i \geq 0$, $\mu_i > 0$ and $\beta_i \in \mathbb{R}$, $i = 1, \ldots, m$. Consider $\beta = (\beta_1, \ldots, \beta_m)$ as a multi-weight and resp. $\lambda = (\lambda_1, \ldots, \lambda_m)$ as a multi-order of fractional integration. The integral operators, defined as follows:

$$I_{(\mu_i),m}^{(\beta_i),(\lambda_i)} f(z) = \begin{cases} \displaystyle\int_0^1 H_{m,m}^{m,0}\left[\tau \,\middle|\, \begin{matrix} \left(\beta_i+\lambda_i+1-\dfrac{1}{\mu_i}, \dfrac{1}{\mu_i}\right)_1^m \\ \left(\beta_i+1-\dfrac{1}{\mu_i}, \dfrac{1}{\mu_i}\right)_1^m \end{matrix}\right] f(z\tau)d\tau, \\ \qquad\qquad\qquad\qquad\qquad\qquad\qquad \text{if } \displaystyle\sum_{i=1}^m \lambda_i > 0, \\ f(z), \quad \text{if } \lambda_1 = \lambda_2 = \cdots = \lambda_m = 0, \end{cases}$$

$$(8.62)$$

are said to be *multiple* (m-*tuple*) *Erdélyi–Kober fractional integration operators* and more generally, the operators of the form:

$$If(z) = z^{\lambda_0} I_{(\mu_i),m}^{(\beta_i),(\lambda_i)} f(z), \quad \text{with } \lambda_0 \geq 0, \qquad (8.63)$$

are briefly called *generalized* (m-*tuple*) *fractional integrals*.

For $m = 1$ the operators (8.62) turn into the *classical Erdélyi–Kober integrals* (8.58), for $m = 2$ these are the *hypergeometric fractional integrals* and for arbitrary $m > 1$ we get the *hyper-Bessel integral operators* as particular cases, see [Kiryakova (1994)]. The theory of GFC based on the operators (8.62) is developed in full detail in Kiryakova's book [Kiryakova (1994)]. Here, we briefly recall some of the basic facts.

Let σ be an arbitrary real number. We denote by $\mathcal{H}(\Omega)$ the space of holomorphic functions in a complex domain Ω, starlike with respect to the origin $z = 0$, and consider the spaces

$$\mathcal{H}_\sigma(\Omega) = \{f(z) = z^p \tilde{f}(z); \ p \geq \sigma, \ \tilde{f}(z) \in \mathcal{H}(\Omega)\}, \quad \mathcal{H}_0(\Omega) := \mathcal{H}(\Omega).$$

For $\beta_i > -1 - \sigma/\mu_i$ and $\lambda_i > 0$, the operators (8.62) map the space $\mathcal{H}_\sigma(\Omega)$ into itself [Kiryakova (1994)].

The main property of the generalized (m-tuple) fractional integrals is that single integrals (8.62), involving H-function, can be equivalently represented by means of commutative compositions of single Erdélyi–Kober integrals of the form (8.58), namely for $f \in \mathcal{H}_\sigma(\Omega)$ with $\beta_i > -1 - \sigma/\mu_i$, $\lambda_i > 0$ and $i = 1, 2, \ldots, m$,

$$I_{(\mu_i),m}^{(\beta_i),(\lambda_i)} f(z) = \prod_{i=1}^{m} I_{\mu_i}^{\beta_i,\lambda_i} f(z)$$

$$= \int_0^1 \cdots \int_0^1 \left[\prod_{i=1}^{m} \frac{(1 - \tau_i)^{\lambda_i - 1} \tau_i^{\beta_i}}{\Gamma(\lambda_i)} \right]$$

$$\times f\left(z\tau_1^{\frac{1}{\mu_1}} \ldots \tau_m^{\frac{1}{\mu_m}} \right) d\tau_1 \ldots d\tau_m. \quad (8.64)$$

If some of the parameters λ_i are zeros, e.g. $\lambda_1 = \cdots = \lambda_k = 0, 1 \leq k \leq m$, the corresponding multipliers $I_{\mu_i}^{\beta_i,\lambda_i} = I$ are identity operators and the multiplicity of (8.62) reduces from m to $m - k$ (same for the order of the kernel H-functions). Decomposition relation (8.64) is the key to various applications of (8.62) and (8.63), arising from the simple but quite effective tools of the G- and H-functions. The generalized fractional derivatives, corresponding to (8.62), are defined in [Kiryakova (1994)] by differ-integral expressions analogously to the idea for (8.53).

Now, by means of analogy with Lemma 8.4, we prove a relation involving the generalized fractional integral (8.62).

Theorem 8.10. *Let* $\omega \in \mathbb{C}$, $\omega \neq 0$, *and* α_i, β_i, $\lambda_i > 0$ *for* $1 \leq i \leq m$. *Then*

$$I_{\left(\frac{1}{\alpha_i}\right),m}^{(\beta_i-1),(\lambda_i)} E_{(\alpha_i),(\beta_i)}^{(\gamma_i),m}(\omega z) = E_{(\alpha_i),(\beta_i+\lambda_i)}^{(\gamma_i),m}(\omega z).$$

Proof. Taking into account the following relations, based on (8.58) and (8.61),

$$\int_0^1 \frac{(1 - \tau_1)^{\lambda_1 - 1} \tau_1^{\beta_1 - 1}}{\Gamma(\lambda_1)} \, E_{(\alpha_i),(\beta_i)}^{(\gamma_i),m} \left(\omega z \tau_1^{\alpha_1} \ldots \tau_m^{\alpha_m} \right) d\tau_1$$

$$= I_{\frac{1}{\alpha_1}}^{\beta_1 - 1, \lambda_1} \, E_{(\alpha_i),(\beta_i)}^{(\gamma_i),m} \left(\omega z \tau_2^{\alpha_2} \ldots \tau_m^{\alpha_m} \right)$$

$$= E_{(\alpha_i),(\beta_1 + \lambda_1, \beta_2, \ldots, \beta_m)}^{(\gamma_i),m} \left(\omega z \tau_2^{\alpha_2} \ldots \tau_m^{\alpha_m} \right),$$

$$\vdots \quad \vdots \quad \vdots$$

$$\int_0^1 \frac{(1 - \tau_m)^{\lambda_m - 1} \tau_m^{\beta_m - 1}}{\Gamma(\lambda_m)} \, E_{(\alpha_i),(\beta_1 + \lambda_1, \ldots, \beta_{m-1} + \lambda_{m-1}, \beta_m)}^{(\gamma_i),m} \left(\omega z \tau_m^{\alpha_m} \right) d\tau_m$$

$$= I_{\frac{1}{\alpha_m}}^{\beta_m - 1, \lambda_m} \, E_{(\alpha_i),(\beta_1 + \lambda_1, \ldots, \beta_{m-1} + \lambda_{m-1}, \beta_m)}^{(\gamma_i),m} (\omega z)$$

$$= E_{(\alpha_i),(\beta_i + \lambda_i)}^{(\gamma_i),m} (\omega z),$$

the subsequent repeated (m times) integrations in (8.64) lead to the identity

$$I_{\left(\frac{1}{\alpha_i} \right),m}^{(\beta_i - 1),(\lambda_i)} \, E_{(\alpha_i),(\beta_i)}^{(\gamma_i),m} (\omega z) = E_{(\alpha_i),(\beta_i + \lambda_i)}^{(\gamma_i),m} (\omega z)$$

$$= \int_0^1 \cdots \int_0^1 \left[\prod_{i=1}^m \frac{(1 - \tau_i)^{\lambda_i - 1} \tau_i^{\beta_i - 1}}{\Gamma(\lambda_i)} \right]$$

$$\times E_{(\alpha_i),(\beta_i)}^{(\gamma_i),m} \left(\omega z \tau_1^{\alpha_1} \ldots \tau_m^{\alpha_m} \right) d\tau_1 \ldots d\tau_m,$$

that proves the theorem. $\square$

Chapter 9

Series in Mittag-Leffler Type Functions

9.1. Multi-index Bessel Series: Theorems of Cauchy–Hadamard and Abel Types

Let $J_n^\mu(z)$, defined by (6.2), and $J_{n-2\lambda,\lambda}^{\mu,m}(z)$, defined by (6.13), be respectively the so-called Bessel–Maitland and generalized Lommel–Wright functions with $z \in \mathbb{C}$. In this chapter, we consider the parameters λ, μ, m and n as follows:

$$\lambda \in \mathbb{C}, \quad \mu > 0, \quad m \in \mathbb{N}, \quad n \in \mathbb{N}_0.$$

Let us recall that for $m = 1$, the Lommel–Wright functions give the 3-index generalized Bessel–Maitland functions. In addition, if $\lambda = 0$ and $\mu = 1$, they are reduced to the Bessel functions $J_n(z)$ of the first type, defined with (1.2). That is why, setting $m = 1$ (respectively $m = 1$, $\lambda = 0$ and $\mu = 1$), all the results, obtained is this chapter, produce the corresponding statements concerning the functions $J_{n-2\lambda,\lambda}^\mu(z)$, respectively $J_n(z)$. In this section, we consider the series in the above-discussed generalized Bessel functions in the complex plane and we briefly call them *multi-index Bessel series* or *Bessel type series*.

In order to specify suitable families of functions for further exposition, we introduce the following denotations:

$$\widetilde{J}_n^\mu(z) = z^n \Gamma(n+1) J_n^\mu(z), \tag{9.1}$$

$$\widetilde{J}^{\mu,m}_{n-2\lambda,\lambda}(z) = (-1)^p z^{-2p} 2^{n+2p} \Gamma^m(\lambda + p + 1)$$
$$\times \Gamma(n - \lambda + p\mu + 1) J^{\mu,m}_{n-2\lambda,\lambda}(z), \qquad (9.2)$$

for $n = 0, 1, 2, \ldots$ and with the corresponding p, defined in Sec. 6.2. Subsequently, for the considerations in this section, we use the series of the kind

$$\sum_{n=0}^{\infty} a_n \widetilde{J}^{\mu}_n(z), \qquad (9.3)$$

as well as

$$\sum_{n=0}^{\infty} a_n \widetilde{J}^{\mu,m}_{n-2\lambda,\lambda}(z), \qquad (9.4)$$

with complex coefficients a_n.

We start with the Cauchy–Hadamard type theorems that give the domains of convergence, i.e. the domains where the series are absolutely convergent but are divergent outside the domains.

Theorem 9.1 (of Cauchy–Hadamard type). *The domain of convergence of the two series* (9.3) *and* (9.4) *is the disk* $D(0; R)$ *with the radius of convergence*

$$R = \left(\limsup_{n \to \infty} (|a_n|)^{\frac{1}{n}} \right)^{-1}. \qquad (9.5)$$

More precisely, both of them are absolutely convergent in the disk $D(0; R)$ *and divergent in the region* $|z| > R$. *The cases* $R = 0$ *and* $R = \infty$ *are included in the general case.*

Proof. Beginning with the series (9.4), we only give the idea of the proof. Namely, using the asymptotic formula (6.17), we evaluate the absolute value of the general term of the series (9.4). Further, the proof proceeds separately for the three cases: $R = \infty, 0 < R < \infty$ and $R = 0$, following the line of Theorem 3.4. We show the absolute convergence of the series (9.4) in the circular domain $D(0; R)$. In the second case, we also prove that the series is divergent for $|z| > R$ and in the third case, it is proven to be divergent for all the complex $z \neq 0$.

The proof of the other case is analogous, but it uses the asymptotic formula (6.15) instead of (6.17). □

The results analogical to the classical Abel's lemma, given below, can be deduced as corollaries from the last theorem.

Corollary 9.1. *Let the series* (9.3) *(respectively* (9.4)) *converge at the point* $z_0 \neq 0$. *Then it is absolutely convergent in the disk* $D(0; |z_0|)$. *Inside the disk* $D(0; R)$, *i.e. on each closed disk* $|z| \leq r$ $(r < R)$, *the convergence is uniform.*

On the boundary $\partial D(0; R)$ of the disk the series may even diverge. Further, the behavior of the considered series 'near' the circle $C(0; R)$ is studied as well. Let $z_0 \in \mathbb{C}$, $0 < R < \infty$, $|z_0| = R$, g_φ be an arbitrary angular domain like (3.4) with a size of $2\varphi < \pi$ and with a vertex at the point $z = z_0$, symmetric with respect to the line through the points 0 and z_0, and d_φ be the part of the domain g_φ, closed between the angle arms and the arc of the circle, centered at the point 0 and touching the arms of the angle (see (3.4) and Figs. 3.1 and 3.2). The following theorem can be formulated for both the series in Bessel–Maitland and generalized Lommel–Wright functions.

Theorem 9.2 (of Abel type). *Let* $\{a_n\}_{n=0}^\infty$ *be a sequence of complex numbers, R be the positive number defined by* (9.5), $F(z)$ *be the sum of any of the series* (9.3), *respectively* (9.4), *in the domain* $D(0; R)$, *i.e.*

$$F(z) = \sum_{n=0}^\infty a_n \tilde{J}_n^\mu(z), \quad z \in D(0; R),$$

respectively

$$F(z) = \sum_{n=0}^\infty a_n \tilde{J}_{n-2\lambda,\lambda}^{\mu,m}(z), \quad z \in D(0; R),$$

and this series converges at the point z_0 of the boundary of $D(0; R)$. Then

(i) *the series* (9.3), *respectively* (9.4), *is uniformly convergent in the domain d_φ and*

(ii) *the following relation holds*:

$$\lim_{z \to z_0} F(z) = \sum_{n=0}^{\infty} a_n \tilde{J}_n^{\mu}(z_0), \tag{9.6}$$

respectively

$$\lim_{z \to z_0} F(z) = \sum_{n=0}^{\infty} a_n \tilde{J}_{n-2\lambda,\lambda}^{\mu,m}(z_0), \tag{9.7}$$

provided $z \in g_{\varphi}$.

Proof. Starting with the series (9.3), we only give the idea of the proof. So, let $z \in d_{\varphi}$. Setting

$$S_k(z) = \sum_{n=0}^{k} a_n \tilde{J}_n^{\mu}(z),$$

we estimate the modulus of the expression

$$S_{k+p}(z) - S_k(z) = \sum_{n=0}^{k+p} a_n \tilde{J}_n^{\mu}(z) - \sum_{n=0}^{k} a_n \tilde{J}_n^{\mu}(z) = \sum_{n=k+1}^{k+p} a_n \tilde{J}_n^{\mu}(z),$$

using essentially the inequality (3.5) and the asymptotic formula (6.15). Following the proof of Theorem 3.5, we prove the uniform convergence. Then the equality (9.6) follows because of the equalities $\lim_{z \to z_0} \tilde{J}_n^{\mu}(z) = \tilde{J}_n^{\mu}(z_0)$ $(n \in \mathbb{N}_0)$.

The proof, referring to the series (9.4), is analogous but it uses the asymptotic formula (6.17). $\square$

Since the proofs of the results given in this section follow the lines of the ones for the Bessel series (Secs. 3.4 and 3.5), the details are omitted. The detailed proofs concerning the series (9.3) can be found in [Paneva-Konovska (2014c)].

9.2. Tauberian and Fatou Type Theorems for Bessel Type Series

In this section, Tauberian type theorems for the summation of divergent series defined by means of the Bessel–Maitland as well as Lommel–Wright functions are given. In order to formulate the results referring to (J, z_0)-summability, defined by Definition 3.6, we first construct suitable families of the kind (3.22). In the beginning, we are going to start with the statements concerning the Bessel–Maitland functions.

Let $z_0 \in \mathbb{C}$, $|z_0| = R$, $0 < R < \infty$, and $J_n^\mu(z_0) \neq 0$ for $n = 0, 1, 2, \ldots$. For the sake of brevity, it is denoted as

$$j_n(z) = J_{n,\mu}^*(z; z_0) = \frac{\widetilde{J}_n^\mu(z)}{\widetilde{J}_n^\mu(z_0)}, \quad (J; z_0) := \{j_n\}_{n \in \mathbb{N}_0}. \tag{9.8}$$

It is evident that all the functions j_n in (9.8) are entire functions and $j_n(z_0) = 1$. Because of that, the family $(J; z_0)$ is of the type defined by (3.22). We also need the series

$$\sum_{n=0}^{\infty} a_n j_n(z) = \sum_{n=0}^{\infty} a_n J_{n,\mu}^*(z; z_0). \tag{9.9}$$

In accordance with Theorem 9.2, the next remark can be made.

Remark 9.1. The (J, z_0)-summation using the system (9.8) is regular, and this property is just a particular case of Theorem 9.2.

So, we get to the following statement, analogical to the classical Tauber theorem.

Theorem 9.3 (of Tauber type). *Let the family $(J; z_0)$ be defined by (9.8). If the numerical series (3.7) is (J, z_0)-summable and*

$$\lim_{n \to \infty} n a_n = 0, \tag{9.10}$$

then it is convergent.

Further, the Littlewood generalization of the just formulated $o(1/n)$ version of Tauber type theorem (Theorem 9.3) is also given, as follows.

Theorem 9.4 (of Littlewood type). *Let the family $(J; z_0)$ be defined by (9.8). If the numerical series (3.7) is (J, z_0)-summable* *and*

$$a_n = O\left(\frac{1}{n}\right), \tag{9.11}$$

then the series (3.7) converges.

Proof. Let z belong to the segment $[0, z_0)$. By using the asymptotic formula (6.15) for the Bessel–Maitland functions, we obtain:

$$a_n J^*_{n,\mu}(z; z_0) = a_n \left(\frac{z}{z_0}\right)^n \frac{1 + \theta_n^\mu(z)}{1 + \theta_n^\mu(z_0)} = a_n \left(\frac{z}{z_0}\right)^n (1 + \widetilde{\theta}_n^\mu(z; z_0)),$$

where $\widetilde{\theta}_n^\mu(z; z_0) = \frac{\theta_n^\mu(z) - \theta_n^\mu(z_0)}{1 + \theta_n^\mu(z_0)}$. Then $\widetilde{\theta}_n^\mu(z; z_0) = O(1/n^\mu)$, due to (6.24) and Remark 6.5 case (iii).

Now, let us write (9.9) in the form:

$$\sum_{n=0}^{\infty} a_n J^*_{n,\mu}(z; z_0) = \sum_{n=0}^{\infty} a_n \left(\frac{z}{z_0}\right)^n (1 + \widetilde{\theta}_n^\mu(z; z_0)). \tag{9.12}$$

Subsequently, denoting $w_n(z) = a_n(z/z_0)^n \widetilde{\theta}_n^\mu(z; z_0)$, we consider the series $\sum_{n=0}^{\infty} w_n(z)$. Since $|w_n(z)| \leq |a_n| |\widetilde{\theta}_n^\mu(z; z_0)|$ and according to the condition (9.11) and estimate (6.24), there exists a constant $\widetilde{C}$, such that $|w_n(z)| \leq \widetilde{C}/n^{1+\mu}$ for $n \in \mathbb{N}$. Since $\sum_{n=1}^{\infty} 1/n^{1+\mu}$ converges, the series $\sum_{n=0}^{\infty} w_n(z)$ is also convergent, even absolutely and uniformly on the segment $[0, z_0]$. Therefore (since $\lim_{z \to z_0} w_n(z) = w_n(z_0) = a_n \widetilde{\theta}_n^\mu(z_0; z_0) = 0$),

$$\lim_{z \to z_0} \sum_{n=0}^{\infty} w_n(z) = \sum_{n=0}^{\infty} \lim_{z \to z_0} w_n(z) = 0.$$

Obviously, the assumption that the series (3.7) is (J, z_0)-summable implies the existence of the limit (3.24). Then, bearing in mind that (9.12) can be written in the form:

$$\sum_{n=0}^{\infty} a_n J^*_{n,\mu}(z; z_0) = \sum_{n=0}^{\infty} a_n \left(\frac{z}{z_0}\right)^n + \sum_{n=0}^{\infty} a_n \left(\frac{z}{z_0}\right)^n \widetilde{\theta}_n^\mu(z; z_0),$$

we conclude that there exists the limit:

$$\lim_{z \to z_0} \sum_{n=0}^{\infty} a_n \left(\frac{z}{z_0} \right)^n \qquad (9.13)$$

and, moreover,

$$\lim_{z \to z_0} \sum_{n=0}^{\infty} a_n J_{n,\mu}^*(z; z_0) = \lim_{z \to z_0} \sum_{n=0}^{\infty} a_n \left(\frac{z}{z_0} \right)^n.$$

From the existence of the limit (9.13), it follows that the series (3.7) is A-summable. Then according to Theorem 3.2, the series (3.7) converges. □

Remark 9.2. Now, the correctness of Theorem 9.3 follows from Theorem 9.4, just as an elementary corollary. It can also be proven independently, and the proof goes analogically to the one of Theorem 9.4.

The results related to the generalized Lommel–Wright functions (6.12) are only formulated, because their proofs go in the same way, but using specifics of the corresponding asymptotic formulae.

Note that each function $\widetilde{J}_{n-2\lambda,\lambda}^{\mu,m}(z)$, $n \in \mathbb{N}_0$, being an entire function, not identically zero, has at most a finite number of zeros in the closed and bounded set $[D(0; R)]$. Moreover, due to Remark 6.8, only a finite number of these functions may have zeros, different from 0.

Let $z_0 \in \mathbb{C}$, $|z_0| = R$, $0 < R < \infty$, and $J_{n-2\lambda,\lambda}^{\mu,m}(z_0) \neq 0$ for $n = 0, 1, 2, \ldots$. For the sake of brevity, we denote

$$j_n(z) = J_{n,\lambda,\mu,m}^*(z; z_0) = \frac{\widetilde{J}_{n-2\lambda,\lambda}^{\mu,m}(z)}{\widetilde{J}_{n-2\lambda,\lambda}^{\mu,m}(z_0)}, \quad (J; z_0) := \{j_n\}_{n \in \mathbb{N}_0}. \qquad (9.14)$$

Obviously, the family $(J; z_0)$ is of the type defined by (3.22), because all the functions j_n in (9.14) are entire functions and $j_n(z_0) = 1$. Now, we formulate Tauberian type theorems by means of the generalized Lommel–Wright functions, using the system (9.14).

Theorem 9.5 (of Tauber type). *Let the system $(J; z_0)$ be defined by (9.14). If the numerical series (3.7) is (J, z_0)-summable and*

$$\lim_{n \to \infty} n a_n = 0, \tag{9.15}$$

then it is convergent.

Theorem 9.6 (of Littlewood type). *Let the system $(J; z_0)$ be defined by (9.14). If the numerical series (3.7) is (J, z_0)-summable and*

$$a_n = O\left(\frac{1}{n}\right), \tag{9.16}$$

then the series (3.7) converges.

Note that this (J, z_0)-summation is regular as well, i.e. the following remark can be made.

Remark 9.3. The considered (J, z_0)-summation is regular, and this property is just a particular case of Theorem 9.2.

Other interesting result, referring to the boundary behavior of the series (9.3) and (9.4), gives a dependence between the regular points of their sums on the boundary and series convergence at such points. The obtained results in this direction for both families are combined in a single theorem as follows.

Theorem 9.7 (of Fatou type). *Let $\{a_n\}_{n=0}^{\infty}$ be a sequence of complex numbers satisfying the conditions*

$$\lim_{n \to \infty} a_n = 0, \quad \limsup_{n \to \infty} (|a_n|)^{\frac{1}{n}} = 1, \tag{9.17}$$

and $F(z)$ be the sum of the series (9.3), respectively (9.4), in the unit disk $D(0; 1)$, i.e.

$$F(z) = \sum_{n=0}^{\infty} a_n \tilde{J}_n^{\mu}(z), \quad z \in D(0; 1), \tag{9.18}$$

respectively

$$F(z) = \sum_{n=0}^{\infty} a_n \tilde{J}_{n,\lambda}^{\mu,m}(z), \quad z \in D(0; 1). \tag{9.19}$$

Let γ be an arbitrary arc of the unit circle $C(0;1)$ with all its points (including the ends) regular to the function F. Then the series (9.3) (respectively (9.4)) converges, even uniformly, on the arc γ.

Proof. We prove the theorem for the series (9.3). Since all the points of the arc γ are regular to the function $F(z)$, there exists a region $G \supset \gamma$ where the function F can be continued. Denoting $\widetilde{G} = G \cup D(0;1)$, we define the function ψ in the region $\widetilde{G}$ by the equality

$$\psi(z) = F(z), \quad z \in D(0;1).$$

More precisely, it means that ψ is a single-valued analytical continuation of F into the domain $\widetilde{G}$.

Let $\rho > 0$ be the distance between the boundary $\partial\widetilde{G}$ of the region $\widetilde{G}$ and the arc γ ($\partial\widetilde{G}$ contains a part of the unit circle $|z| = 1$), and take the points ζ_1 and ζ_2 as

$$\zeta_1, \zeta_2 \notin \gamma, \quad |\zeta_1| = |\zeta_2| = 1,$$

such that the distances between each of the points ζ_1 and ζ_2 and the respective closer end of the arc γ are equal to $\rho/2$, and denote (see Fig. 3.3)

$$z_1 = \zeta_1\left(1 + \frac{\rho}{2}\right), \quad z_2 = \zeta_2\left(1 + \frac{\rho}{2}\right).$$

Define the auxiliary function

$$\varphi_n(z) = \psi(z) - \sum_{k=0}^{n} a_k \widetilde{J}_k^\mu(z) \tag{9.20}$$

and note that, according to Remark 6.8, there exists a natural number N_0 such that $\widetilde{J}_n^\mu(z) \neq 0$ when z belongs to a nonempty compact subset of $\mathbb{C}$, $z \neq 0$ and $n > N_0$. Now, letting $n \geq N_0$, we introduce the notation:

$$\omega_n(z) = \frac{\varphi_n(z)}{\widetilde{J}_{n+1}^\mu(z)}(z - \zeta_1)(z - \zeta_2), \quad z \neq 0; \quad \omega_n(0) = a_{n+1}\zeta_1\zeta_2.$$

$$\tag{9.21}$$

In order to prove that the sequence $\left\{\sum_{k=0}^n a_k \widetilde{J}_k^\mu(z)\right\}$ is uniformly convergent on the arc γ, it is sufficient to show that the sequence

$\{\omega_n(z)\}_{n=N_0}^{\infty}$ tends uniformly to zero on the boundary $\partial\Delta$ of the sector $\Delta = Oz_1z_2$ and then estimate $\varphi_n(z)$ on the arc γ.

To this end, we come back to (6.24) and the Γ quotient property, given in Remark 6.5 case (iii). Let us just mention that since $\lim_{n\to\infty} \frac{1}{n^\mu} = 0$, there exist numbers C and $\tilde{N} > N_0$ such that $\left|1+\theta_n^\mu(z)\right| \leq C/2$ for all the values of $n \in \mathbb{N}$ and $1/2 \leq \left|1+\theta_n^\mu(z)\right| \leq 2$ for $n > \tilde{N}$ on an arbitrary compact subset of $\mathbb{C}$.

Now, taking $\varepsilon > 0$ and setting

$$R = 1 + \frac{\rho}{2}, \quad \varepsilon_1 = \frac{\varepsilon\rho^3}{8(8CR^2 + \rho)},$$

$$M = \max_{z\in[\Delta]} |\psi(z)| \quad ([\Delta] = \Delta \cup \partial\Delta),$$

we have to consider four cases as follows.

(i) First, let $z \in (O, \zeta_1) \cup (O, \zeta_2) \subset D(0; 1)$.

In the unit disk, according to (9.20), we have consecutively:

$$\omega_n(z) = \sum_{k=0}^{\infty} a_{n+k+1} \frac{\widetilde{J}_{n+k+1}^\mu(z)}{\widetilde{J}_{n+1}^\mu(z)}(z - \zeta_1)(z - \zeta_2),$$

$$\omega_n(z) = \sum_{k=0}^{\infty} a_{n+k+1} z^k \frac{(1 + \theta_{n+k+1}^\mu(z))}{(1 + \theta_{n+1}^\mu(z))}(z - \zeta_1)(z - \zeta_2).$$

$$(9.22)$$

Since both $a_n \to 0$ and $|(z - \zeta_1)(z - \zeta_2)| < 2(1 - |z|)$, there exists a number $N_1 = N_1(\varepsilon_1) > \tilde{N}$, such that

$$|\omega_n(z)| < \varepsilon_1 \sum_{k=0}^{\infty} |z|^k \left| \frac{(1 + \theta_{n+k+1}^\mu(z))}{(1 + \theta_{n+1}^\mu(z))} \right| |(z - \zeta_1)||(z - \zeta_2)|$$

$$< 2C\varepsilon_1 \sum_{k=0}^{\infty} |z|^k (1 - |z|) = 2C\varepsilon_1$$

for $n > N_1$, i.e.

$$|\omega_n(z)| < 2C\varepsilon_1. \tag{9.23}$$

(ii) Let $z \in (\zeta_1, z_1) \cup (\zeta_2, z_2)$.

In this case, $|z-\zeta_1| = |z|-1$, $|z-\zeta_2| \le |z|+|\zeta_2| < 2R$, and taking into account Theorem 6.4, the asymptotic formulae (6.15) and (9.20), we can write the following inequalities for the absolute value of $\omega_n(z)$:

$$\omega_n(z) = \frac{\psi(z) - \sum_{k=0}^{n} a_k \, z^k \left(1 + \theta_k^\mu(z)\right)}{z^{n+1}\left(1 + \theta_{n+1}^\mu(z)\right)}(z - \zeta_1)(z - \zeta_2),$$

namely

$$|\omega_n(z)| \le \frac{M + \sum_{k=0}^{n} |a_k||z|^k \left|\left(1 + \theta_k^\mu(z)\right)\right|}{|z|^{n+1}\left|\left(1 + \theta_{n+1}^\mu(z)\right)\right|} \, 2R(|z| - 1)$$

$$< 2R\left(2M + \sum_{k=0}^{N_1} C|a_k|R^k\right) \frac{(|z| - 1)}{|z|^{n+1}}$$

$$+ 2\varepsilon_1 RC \frac{(|z| - 1)}{|z|^{n+1}} \sum_{k=N_1+1}^{n} |z|^k.$$

Furthermore, bearing in mind that on the one hand,

$$\frac{(|z| - 1)}{|z|^{n+1}} < \frac{(|z| - 1)}{|z|^{n+1} - 1} = \frac{1}{|z|^n + \cdots + 1} < \frac{1}{n + 1},$$

and on the other hand;

$$\sum_{k=N_1+1}^{n} |z|^k = \frac{|z|^{n+1} - |z|^{N_1+1}}{(|z| - 1)} < \frac{|z|^{n+1}}{(|z| - 1)},$$

we conclude that

$$|\omega_n(z)| < \frac{2R}{n + 1}\left(2M + \sum_{k=0}^{N_1} C|a_k|R^k\right) + 2\varepsilon_1 RC.$$

Then, since $n^{-1} \to 0$, there exists a number $N_2 = N_2(\varepsilon_1) > N_1$ such that

$$\frac{2R}{n + 1}\left(2M + \sum_{k=0}^{N_1} C|a_k|R^k\right) < \varepsilon_1$$

for $n > N_2$, i.e.

$$|\omega_n(z)| < (1 + 2RC)\varepsilon_1. \tag{9.24}$$

(iii) z belong to the arc $\widehat{z_1 z_2}$ (including the ends).
Then $|z - \zeta_1| < 2R$, $|z - \zeta_2| < 2R$ and hence

$$|\omega_n(z)| < \frac{4R^2 \left(2M + \sum_{k=0}^n C|a_k|R^k\right)}{R^{n+1}}$$

$$< \frac{4\left(2M + \sum_{k=0}^{N_1} C|a_k|R^k\right)}{R^{n-1}} + \frac{8\varepsilon_1 CR^2}{\rho}.$$

Since $R^{-n} \to 0$, there exists a number $N_3 = N_3(\varepsilon_1) > N_1$, such that

$$|\omega_n(z)| < \left(\frac{8CR^2}{\rho} + 1\right)\varepsilon_1 \tag{9.25}$$

for $n > N_3$.

(iv) Let $z \in \{O, \zeta_1, \zeta_2\}$.
In this case, (9.21) produces $\omega_n(0) = a_{n+1}\zeta_1\zeta_2$, from which

$$|\omega_n(0)| = |a_{n+1}| < \varepsilon_1 \text{ for } n > N_1 \quad \text{and} \quad \omega_n(\zeta_{1,2}) = 0. \tag{9.26}$$

Let $N = \max\{N_1, N_2, N_3\}$ and $n > N$, then bearing in mind the inequalities (9.23)–(9.26), we can write the boundary of the region Δ as

$$|\omega_n(z)| < \max\left(2C\varepsilon_1, (2RC + 1)\varepsilon_1, \left(\frac{8CR^2}{\rho} + 1\right)\varepsilon_1\right)$$

$$= \left(\frac{8CR^2}{\rho} + 1\right)\varepsilon_1.$$

Hence, according to the maximum modulus principle, the following inequality can be deduced:

$$|\omega_n(z)| < \left(\frac{8CR^2 + \rho}{\rho}\right)\varepsilon_1, \quad z \in \gamma. \tag{9.27}$$

Eventually, according to (6.15), (9.20) and (9.22), since $|z| = 1$ on the arc γ,

$$|\omega_n(z)| = \frac{\left|\psi(z) - \sum_{k=0}^{n} a_k \widetilde{J}_k^{\mu}(z)\right|}{|z^{n+1}| \left|1 + \theta_{n+1}^{\mu}(z)\right|} |z - \zeta_1||z - \zeta_2|$$

$$> \frac{\rho^2}{8} \left|\psi(z) - \sum_{k=0}^{n} a_k \widetilde{J}_k^{\mu}(z)\right|, \qquad (9.28)$$

from which, by applying the inequality (9.27), the relation (9.28) yields

$$\left|\psi(z) - \sum_{k=0}^{n} a_k \widetilde{J}_k^{\mu}(z)\right| < \frac{8}{\rho^2} |\omega_n(z)| < \frac{8\varepsilon_1}{\rho^3} \left(8CR^2 + \rho\right) = \varepsilon, \quad z \in \gamma,$$

that proves the theorem in this case. The proof for the series (9.4) goes analogously. $\qquad \square$

9.3. Mittag-Leffler and Generalized Mittag-Leffler Series: Cauchy–Hadamard and Abel Type Theorems

Let $E_n(z)$, $E_{\alpha,n}(z)$, defined by (7.1) and (7.2), $E_{\alpha,n}^{\gamma}(z)$, defined by (7.8), be respectively the Mittag-Leffler functions and their 3-index generalizations with $z \in \mathbb{C}$ and the parameters α, $\gamma \in \mathbb{C}$, $n \in \mathbb{N}$ and $\Re(\alpha) > 0$. Let us remind that for $\gamma = 1$, the generalized Mittag-Leffler functions coincide with the Mittag-Leffler functions. Because of that, setting $\gamma = 1$, all the results obtained in this chapter produce the corresponding statements concerning the functions $E_{\alpha,n}(z)$.

In what follows, we use the notations:

$$\widetilde{E}_0(z) = 1; \quad \widetilde{E}_n(z) = z^n E_n(z), \quad n \in \mathbb{N};$$

$$\widetilde{E}_{\alpha,0}^0(z) = 1; \quad \widetilde{E}_{\alpha,n}^0(z) = \Gamma(n) \, z^n E_{\alpha,n}^0(z), \quad n \in \mathbb{N}; \qquad (9.29)$$

$$\widetilde{E}_{\alpha,n}^{\gamma}(z) = \frac{\Gamma(\alpha p + n)}{(\gamma)_p} \, z^{n-p} E_{\alpha,n}^{\gamma}(z), \quad \gamma \neq 0, \; n \in \mathbb{N}_0,$$

with the corresponding p, determined in Sec. 7.4 (here, $\widetilde{E}_0(z) = 1$ and $\widetilde{E}_{\alpha,0}^0(z) = 1$ are introduced just for completeness).

In this section, we consider the series in such a type of functions in the complex plane, namely:

$$\sum_{n=0}^{\infty} a_n \widetilde{E}_n(z), \tag{9.30}$$

respectively

$$\sum_{n=0}^{\infty} a_n \widetilde{E}_{\alpha,n}^{\gamma}(z), \tag{9.31}$$

with complex coefficients a_n ($n = 0, 1, 2, \ldots$). The first of them we briefly call the *Mittag-Leffler series* and second one as the *generalized Mittag-Leffler series*, or both series are simply called the *Mittag-leffler type series*. Since the proofs of the assertions in this section follow the lines of the ones referring to the Bessel functions, the details are omitted here. First, we state a theorem of Cauchy–Hadamard type for the series (9.30) and (9.31).

Theorem 9.8 (of Cauchy–Hadamard type). *The domain of convergence of each of the series* (9.30) *and* (9.31) *with complex coefficients a_n is the disk $D(0; R)$ with a radius of convergence*

$$R = \left(\limsup_{n \to \infty} (|a_n|)^{\frac{1}{n}} \right)^{-1}. \tag{9.32}$$

More precisely, both series are absolutely convergent in the disk $D(0; R)$ and divergent in the domain $|z| > R$. The cases $R = 0$ and $R = \infty$ fall in the general case.

Corollary 9.2. *Let any of the series* (9.30) *and* (9.31) *converge at the point $z_0 \neq 0$. Then it is absolutely convergent in the disk $D(0; |z_0|)$. Moreover, inside the disk $D(0; R)$, i.e. on each closed disk $|z| \leq r < R$ (R defined by* (9.32)*), both series are uniformly convergent.*

Let $z_0 \in \mathbb{C}$, $0 < R < \infty$, $|z_0| = R$, g_φ be an arbitrary angular domain like (3.4) with a size of $2\varphi < \pi$ and with a vertex at the point $z = z_0$, and d_φ be the part of g_φ defined by (3.4). The next theorem refers to the uniform convergence of the series (9.30) and (9.31) in the set d_φ and the existence of the limits of their sums at the point z_0, provided $|z| < R$ and $z \in g_\varphi$.

Theorem 9.9 (of Abel type). *Let $\{a_n\}_{n=0}^\infty$ be a sequence of complex numbers, R be the real number defined by (9.32), $0 < R < \infty$, and $F(z)$, and $G(z)$ be the sums of the series (9.30) and (9.31) in the domain $D(0; R)$, i.e.*

$$F(z) = \sum_{n=0}^\infty a_n \widetilde{E}_n(z), \quad z \in D(0; R), \tag{9.33}$$

$$G(z) = \sum_{n=0}^\infty a_n \widetilde{E}_{\alpha,n}^\gamma(z), \quad z \in D(0; R). \tag{9.34}$$

If any of these series ((9.30) or respectively (9.31)) converges at the point z_0 of the boundary $\partial D(0; R) = C(0; R)$, then

(i) *the series (9.30), respectively (9.31), is uniformly convergent in the domain d_φ and*
(ii) *the following relation holds:*

$$\lim_{z \to z_0} F(z) = \sum_{n=0}^\infty a_n \widetilde{E}_n(z_0), \tag{9.35}$$

respectively

$$\lim_{z \to z_0} F(z) = \sum_{n=0}^\infty a_n \widetilde{E}_{\alpha,n}^\gamma(z_0), \tag{9.36}$$

provided $z \in g_\varphi$.

The details of the proofs referring to the series (9.31) and the equality (9.36) can be seen in [Paneva-Konovska (2014b)], except for the uniformity and Corollary 9.2. The ideas of the last ones follow analogously the work [Paneva-Konovska (2012b)], where the proofs of the corresponding results concerning the series (9.30) and equality (9.35) are given, and also analogously to Secs. 3.4 and 3.5 for the corresponding statements relating to the Bessel functions series.

9.4. Tauberian and Fatou Type Theorems for Mittag-Leffler Type Series

Let $\widetilde{E}_n$ and $\widetilde{E}_{\alpha,n}^{\gamma}$ be the functions defined by the relations (9.29), with α, $\gamma \in \mathbb{C}$ and $\Re(\alpha) > 0$. Let $z_0 \in \mathbb{C}$, $|z_0| = R$ and $0 < R < \infty$, $\widetilde{E}_n(z_0) \neq 0$ and $\widetilde{E}_{\alpha,n}^{\gamma}(z_0) \neq 0$ as well (the last two conditions can be written because both the functions $\widetilde{E}_n$ and $\widetilde{E}_{\alpha,n}^{\gamma}$ are entire functions, and additionally keeping in view the Remarks 7.10 and 7.11). For convenience, denote

$$j_n(z) := E_n^*(z; z_0) = \frac{\widetilde{E}_n(z)}{\widetilde{E}_n(z_0)}, \quad (J; z_0) := \{j_n\}_{n \in \mathbb{N}_0}, \qquad (9.37)$$

respectively

$$j_n(z) := E_{\alpha,n,\gamma}^*(z; z_0) = \frac{\widetilde{E}_{\alpha,n}^{\gamma}(z)}{\widetilde{E}_{\alpha,n}^{\gamma}(z_0)}, \quad (J; z_0) := \{j_n\}_{n \in \mathbb{N}_0}. \qquad (9.38)$$

It is clear that all the functions j_n in formulae (9.37) and (9.38) are entire functions satisfying the condition $j_n(z_0) = 1$.

In order to give the further results, we consider the corresponding series of the form (3.23), i.e.

$$\sum_{n=0}^{\infty} a_n j_n(z) = \sum_{n=0}^{\infty} a_n E_n^*(z; z_0), \qquad (9.39)$$

respectively

$$\sum_{n=0}^{\infty} a_n j_n(z) = \sum_{n=0}^{\infty} a_n E_{\alpha,n,\gamma}^*(z; z_0), \qquad (9.40)$$

used in Definition 3.6 of (J, z_0)-summation of a given numerical series. Let us point out that according to Theorem 9.9, the following remark, referring to such a kind of summation, can be, made.

Remark 9.4. The (J, z_0)-summation of the numerical series (3.7) using the family (9.37), respectively (9.38), is regular and this property is just a particular case of Theorem 9.9.

Theorem 9.10 (of Tauber type). *If* $\{a_n\}_{n=0}^{\infty}$ *is a sequence of complex numbers with*

$$\lim_{n\to\infty} na_n = 0 \qquad (9.41)$$

and the numerical series (3.7) *is* (J, z_0)-*summable (by means of the system* (9.37), *respectively* (9.38)), *then the series* (3.7) *is summable.*

Proof. We start with the summability in a sense of the family (9.38) of Prabhakar's generalizations of the Mittag-Leffler functions. In the case $\gamma = 0$, the proof follows immediately because, according to Lemma 7.12, the series (9.40) reduces to a power series. Let now $\gamma \neq 0$ and z belong to the segment $[0, z_0)$. Taking into account the asymptotic formula (7.29) for the generalized Mittag-Leffler functions, we obtain

$$a_n E_{\alpha,n,\gamma}^{*}(z; z_0) = a_n \left(\frac{z}{z_0}\right)^n \left(1 + \widetilde{\theta}_n(z; z_0)\right),$$

where $\widetilde{\theta}_n(z; z_0) = \frac{\theta_{\alpha,n}^{\gamma}(z) - \theta_{\alpha,n}^{\gamma}(z_0)}{1 + \theta_{\alpha,n}^{\gamma}(z_0)}$. Then, due to (7.30),

$$\left|\widetilde{\theta}_n(z; z_0)\right| = O\left(\frac{1}{n^{\Re(\alpha)}}\right). \qquad (9.42)$$

Writing (9.40) in the following way:

$$\sum_{n=0}^{\infty} a_n E_{\alpha,n,\gamma}^{*}(z; z_0) = \sum_{n=0}^{\infty} a_n \left(\frac{z}{z_0}\right)^n \left(1 + \widetilde{\theta}_n(z; z_0)\right), \qquad (9.43)$$

we consider the series $\sum_{n=0}^{\infty} w_n(z)$ with $w_n(z) = a_n (z/z_0)^n \, \widetilde{\theta}_n(z; z_0)$. Since $|w_n(z)| \leq |a_n| \, |\widetilde{\theta}_n(z; z_0)|$ and according to the condition (9.41) and the relationship (9.42), there exists a constant C, such that $|w_n(z)| \leq C/n^{1+\Re(\alpha)}$. Further, since $\Re(\alpha) > 0$, the numerical series $\sum_{n=1}^{\infty} 1/n^{1+\Re(\alpha)}$ converges, from which the series $\sum_{n=0}^{\infty} w_n(z)$ is also convergent, even absolutely and uniformly on the segment $[0, z_0)$. Therefore (since $\lim_{z\to z_0} w_n(z) = 0$),

$$\lim_{z\to z_0} \sum_{n=0}^{\infty} w_n(z) = \sum_{n=0}^{\infty} \lim_{z\to z_0} w_n(z) = 0.$$

Obviously, the assumption that the series (3.7) is (J, z_0)-summable implies the existence of the limit (3.24). Then, bearing in mind that (9.43) can be written in the form:

$$\sum_{n=0}^{\infty} a_n E_{\alpha,n,\gamma}^*(z; z_0) = \sum_{n=0}^{\infty} a_n \left(\frac{z}{z_0}\right)^n + \sum_{n=0}^{\infty} w_n(z),$$

we conclude that there exists the limit

$$\lim_{z \to z_0} \sum_{n=0}^{\infty} a_n \left(\frac{z}{z_0}\right)^n \qquad (9.44)$$

and, moreover,

$$\lim_{z \to z_0} \sum_{n=0}^{\infty} a_n E_{\alpha,n,\gamma}^*(z; z_0) = \lim_{z \to z_0} \sum_{n=0}^{\infty} a_n \left(\frac{z}{z_0}\right)^n.$$

From the existence of the limit (9.44), it follows that the series (3.7) is A-summable. Then according to Theorem 3.1, the series (3.7) converges, which proves the theorem in this case.

The summability referring to the system (9.37) can be proved analogously, using the series (9.39) instead of (9.40) and the corresponding asymptotic formula. □

A Littlewood generalization of the $o(1/n)$ version of the Tauber type theorem (Theorem 9.10) is given below. Since its proof follows the lines of the one of Theorem 9.10, only the result is formulated.

Theorem 9.11 (of Littlewood type). *If $\{a_n\}_{n=0}^{\infty}$ is a sequence of complex numbers with $a_n = O(1/n)$ and the numerical series, given by (3.7), is (J, z_0)-summable (by means of the system (9.37), respectively (9.38)), then series is summable.*

Next result set out below is the Fatou type theorem, connected with the considered type of series.

Theorem 9.12 (of Fatou type). *Let $\{a_n\}_{n=0}^{\infty}$ be a sequence of complex numbers satisfying the conditions*

$$\lim_{n \to \infty} a_n = 0, \quad \limsup_{n \to \infty} (|a_n|)^{\frac{1}{n}} = 1,$$

and $F(z)$ be the sum of the series (9.30), *respectively* (9.31), *in the unit disk $D(0; 1)$, i.e.*

$$F(z) = \sum_{n=0}^{\infty} a_n \widetilde{E}_n(z), \quad z \in D(0; 1), \tag{9.45}$$

respectively

$$F(z) = \sum_{n=0}^{\infty} a_n \widetilde{E}_{\alpha,n}^{\gamma}(z), \quad z \in D(0; 1). \tag{9.46}$$

Let γ be an arbitrary arc of the unit circle $C(0; 1)$ with all its points (including the ends) regular to the function F. Then the series (9.30) *(respectively* (9.31)*) converges, even uniformly, on the arc γ.*

Proof. We simply give an idea of the proof. Because of the regularity, there exists a region $G \supset \gamma$ where the function F can be continued. Denoting $\widetilde{G} = G \cup D(0; 1)$, we define the function ψ in the region $\widetilde{G}$ by the equality

$$\psi(z) = F(z), \quad z \in D(0; 1),$$

i.e. it means that F has a single-valued analytical continuation into $\widetilde{G}$.

Let $\rho > 0$ be the distance between the boundary $\partial \widetilde{G}$ of the region $\widetilde{G}$ and the arc γ ($\partial \widetilde{G}$ contains a part of the unit circle $C(0; 1)$), and take the points $\zeta_1, \zeta_2 \notin \gamma$, $|\zeta_1| = |\zeta_2| = 1$ (see Fig. 3.3), such that the distances between each of the points ζ_1 and ζ_2 and the respective proximal end of the arc γ are equal to $\rho/2$, also $z_1 = \zeta_1(1 + \rho/2)$ and $z_2 = \zeta_2(1 + \rho/2)$.

We define the auxiliary functions:

$$\varphi_n(z) = \psi(z) - \sum_{k=0}^{n} a_k \widetilde{E}_{\alpha,k}^{\gamma}(z),$$

and point out that according to Remark 7.11, there exists a natural number N_0 such that $\widetilde{E}_{\alpha,n}^{\gamma}(z) \neq 0$ in the bounded set $[\widetilde{G}]$ when $z \neq 0$ and $n > N_0$. Now, letting $n \geq N_0$, we introduce the notation:

$$\omega_n(z) = \frac{\varphi_n(z)}{\widetilde{E}_{\alpha,n+1}^{\gamma}(z)}(z - \zeta_1)(z - \zeta_2), \quad \omega_n(0) = a_{n+1}\zeta_1\zeta_2.$$

The proof continues in the same way as in Theorem 9.7, showing that the sequence $\{\omega_n(z)\}_{n=0}^{\infty}$ tends uniformly to zero on the boundary $\partial\Delta$ of the sector $\Delta = Oz_1z_2$, and then using this fact to prove that the sequence $\left\{\sum_{k=0}^{n} a_k \widetilde{E}_{\alpha,k}^{\gamma}(z)\right\}$ is uniformly convergent on the arc γ. The details are omitted, but they can be seen, for example, in [Paneva-Konovska (2014d)].

The proof for the other series goes analogously.   □

9.5. Multi-index Mittag-Leffler Series: Cauchy–Hadamard and Abel Type Theorems

Let $E_{(\alpha_i),(\mu_{i_0}(n))}(z)$, defined by (8.5), and $J_{(\nu_{i_0}(n))}^{(m)}(z)$, defined by (6.20), be respectively the multi-index Mittag-Leffler functions and hyper-Bessel functions with $z \in \mathbb{C}$, respectively $z \in \mathbb{C}\backslash(-\infty, 0]$, and the parameters $\alpha_i > 0$, $\mu_i, \nu_i \in \mathbb{C}$, $\Re(\nu_i + 1) > 0$ ($i = 1, \ldots, m$) and $n \in \mathbb{N}_0$.

To specify a suitable family of functions for further exposition, we modify a little bit the functions, mentioned above, by multiplying with appropriate coefficients and power functions.

Let $\widetilde{E}_{(\alpha_i),(\mu_{i_0}(n))}$ be the function:

$$\widetilde{E}_{(\alpha_i),(\mu_{i_0}(n))}(z) = z^{n-p} E_{(\alpha_i),(\mu_{i_0}(n))}(z)\Gamma(\alpha_{i_0}p + n)\prod_{i=1}^{m}{}'\Gamma(\alpha_i p + \mu_i),$$

$$(9.47)$$

with the corresponding p, determined in Sec. 8.2 and $\widetilde{J}_{(\nu_{i_0}(n))}^{(m)}$ be defined by

$$\widetilde{J}_{(\nu_{i_0}(n))}^{(m)}(z) = C_{m,n}\left(\frac{z}{m+1}\right)^{-\sum_{i=1}^{m}{}'\nu_i} J_{(\nu_{i_0}(n))}^{(m)}(z), \qquad (9.48)$$

with

$$C_{m,n} = (m+1)^n\,\Gamma(n+1)\prod_{i=1}^{m}{}'\Gamma(\nu_i + 1). \qquad (9.49)$$

In this section, the series in such a type of entire functions are considered, respectively of the form:

$$\sum_{n=0}^{\infty} a_n \widetilde{E}_{(\alpha_i),(\mu_{i_0}(n))}(z) \tag{9.50}$$

as well as

$$\sum_{n=0}^{\infty} a_n \widetilde{J}_{(\nu_{i_0}(n))}^{(m)}(z), \tag{9.51}$$

with complex coefficients a_n $(n = 0, 1, 2, \ldots)$ and we study their convergence in the complex plane $\mathbb{C}$. Their disks of convergence are found and the behavior 'near' the boundaries is studied, by means of theorems of Cauchy–Hadamard and Abel types. Since the proofs of the results given in the present and the following sections follow the lines of the ones for the Bessel series (Secs. 3.4–3.8) and those in the preceding Secs. 9.1–9.4 as well, the details are omitted.

First, we state a theorem of the Cauchy–Hadamard type.

Theorem 9.13 (of Cauchy–Hadamard type). *The domain of convergence of the series* (9.50) *and* (9.51) *with complex coefficients* a_n *is the disk* $D(0; R)$ *with a radius of convergence*

$$R = \left(\limsup_{n \to \infty} (|a_n|)^{\frac{1}{n}} \right)^{-1}. \tag{9.52}$$

More precisely, both series (9.50) *and* (9.51) *are absolutely convergent in the disk* $D(0; R)$ *and divergent in the domain* $|z| > R$. *The cases* $R = 0$ *and* $R = \infty$ *fall in the general case.*

The following statement can be stated as an elementary corollary of this theorem.

Corollary 9.3. *Let any of the series* (9.50) *and* (9.51) *converge at the point* $z_0 \neq 0$. *Then it is absolutely convergent in the disk* $D(0; |z_0|) \subset \mathbb{C}$. *Inside the disk* $D(0; R)$, *i.e. on each closed disk* $|z| \leq r < R$ $(R$ *defined by* (9.52)$)$, *the convergence is uniform.*

Let $z_0 \in \mathbb{C}$, $0 < R < \infty$, $|z_0| = R$ and g_φ be an arbitrary angular domain like (3.4) with a size of $2\varphi < \pi$ and with a vertex at the point $z = z_0$. Also, let d_φ be the part of the angular domain g_φ, located between the angle arms and the arc of the circle centered at the point 0 and touching the arms of the angle. The next theorem refers to the uniform convergence of the series (9.50) and (9.51) in the set d_φ and the existence of the limits of their sums at the point z_0, provided $|z| < R$ and $z \in g_\varphi$.

Theorem 9.14 (of Abel type). *Let* $\{a_n\}_{n=0}^\infty$ *be a sequence of complex numbers, R be the real number defined by (9.52), $0 < R < \infty$, and $F(z)$ and $G(z)$ be the sums of the series (9.50) and, respectively, (9.51), in the domain $D(0; R)$. If any of these series ((9.50), respectively (9.51)) converges at the point z_0 of the boundary of $D(0; R)$, then it is uniformly convergent in the domain d_φ and the following relation holds:*

$$\lim_{z \to z_0} F(z) = \sum_{n=0}^\infty a_n \widetilde{E}_{(\alpha_i),(\mu_{i_0}(n))}(z_0), \tag{9.53}$$

respectively

$$\lim_{z \to z_0} G(z) = \sum_{n=0}^\infty a_n \widetilde{J}_{(\nu_{i_0}(n))}^{(m)}(z_0), \tag{9.54}$$

provided $z \in g_\varphi$.

Note that the proofs of the results in this section go analogously to those in Secs. 3.4 and 3.5. For this reason the details are omitted. However, the proofs concerning the series (9.50) and the equality (9.53) can be seen in [Paneva-Konovska (2012c)], excepting the uniformity and Corollary 9.3 that are published in [Paneva-Konovska (2015a)]. The corresponding results for the series (9.51) are considered in [Paneva-Konovska (2014a)] where the ideas of their proofs are given.

9.6. Tauberian and Fatou Type Theorems for Multi-index Mittag-Leffler Series

Let $\widetilde{E}_{(\alpha_i),(\mu_{i_0}(n))}(z)$ and $\widetilde{J}^{(m)}_{(\nu_{i_0}(n))}(z)$ be the functions defined by the relation (9.47), respectively (9.48), $z_0 \in \mathbb{C}$ with $|z_0| = R$, $0 < R < \infty$, $\widetilde{E}_{(\alpha_i),(\mu_{i_0}(n))}(z_0) \neq 0$ and $\widetilde{J}^{(m)}_{(\nu_{i_0}(n))}(z_0) \neq 0$ (the last two conditions can be written because the functions $\widetilde{E}_{(\alpha_i),(\mu_{i_0}(n))}(z)$ and $\widetilde{J}^{(m)}_{(\nu_{i_0}(n))}(z)$ are entire functions, and additionally keeping in view Remarks 8.4 and 6.8). For convenience, denote

$$j_n(z) := E^{\star}_{(\alpha_i),(\mu_{i_0}(n))}(z; z_0) = \frac{\widetilde{E}_{(\alpha_i),(\mu_{i_0}(n))}(z)}{\widetilde{E}_{(\alpha_i),(\mu_{i_0}(n))}(z_0)}, \quad n \in \mathbb{N}_0, \quad (9.55)$$

respectively

$$j_n(z) := J^{\star(m)}_{(\nu_{i_0}(n))}(z; z_0) = \frac{\widetilde{J}^{(m)}_{(\nu_{i_0}(n))}(z)}{\widetilde{J}^{(m)}_{(\nu_{i_0}(n))}(z_0)}, \quad n \in \mathbb{N}_0. \quad (9.56)$$

It is evident that all the functions j_n in formulae (9.55) and (9.56) are entire functions satisfying the condition $j_n(z_0) = 1$, i.e. the corresponding families $(J; z_0) := \{j_n\}_{n \in \mathbb{N}_0}$ are of the kind (3.22).

In order to provide the further results, we consider the series of the form (3.23), i.e.

$$\sum_{n=0}^{\infty} a_n j_n(z) = \sum_{n=0}^{\infty} a_n E^{\star}_{(\alpha_i),(\mu_{i_0}(n))}(z; z_0), \quad (9.57)$$

respectively

$$\sum_{n=0}^{\infty} a_n j_n(z) = \sum_{n=0}^{\infty} a_n J^{\star(m)}_{(\nu_{i_0}(n))}(z; z_0), \quad (9.58)$$

like in Definition 3.6 of (J, z_0)-summation of a given numerical series. Taking into consideration Theorem 9.14, the following remark, referring to such a kind of summation, can be made.

Remark 9.5. The (J, z_0)-summation of the numerical series (3.7), by means of the family (9.55), respectively (9.56), is regular and this property is just a particular case of Theorem 9.14.

Further, the following statement, which is the inverse of Theorem 9.14, holds true.

Theorem 9.15 (of Tauber type). *If $\{a_n\}_{n=0}^{\infty}$ is a sequence of complex numbers with $\lim_{n\to\infty} na_n = 0$ and the numerical series (3.7) is (J, z_0)-summable (in a sense of the functions (9.55), respectively (9.56)), then the series (3.7) is summable.*

A Littlewood generalization of this theorem is given below.

Theorem 9.16 (of Littlewood type). *If $\{a_n\}_{n=0}^{\infty}$ is a sequence of complex numbers with $a_n = O(1/n)$ and the numerical series, given in (3.7), is (J, z_0)-summable (in a sense of the functions (9.55), respectively (9.56)), then it is summable.*

In the end of this section, the Fatou type theorem is formulated.

Theorem 9.17 (of Fatou type). *Let $\{a_n\}_{n=0}^{\infty}$ be a sequence of complex numbers satisfying the conditions*

$$\lim_{n\to\infty} a_n = 0, \quad \limsup_{n\to\infty} (|a_n|)^{\frac{1}{n}} = 1,$$

and $F(z)$ be the sum of the series (9.50), or respectively (9.51), in the unit disk $D(0;1)$, i.e.

$$F(z) = \sum_{n=0}^{\infty} a_n \widetilde{E}_{(\alpha_i),(\mu_{i_0}(n))}(z), \quad z \in D(0;1), \qquad (9.59)$$

or respectively

$$F(z) = \sum_{n=0}^{\infty} a_n \widetilde{J}^{(m)}_{(\nu_{i_0}(n))}(z), \quad z \in D(0;1). \qquad (9.60)$$

Let γ be an arbitrary arc of the unit circle $C(0;1)$ with all its points (including the ends of the arc) regular to the function F. Then the series (9.50) (respectively (9.51)) converges, even uniformly, on the arc γ.

Note that the proofs of the Tauberian type theorems in a sense of the functions (9.55) are given in [Paneva-Konovska (2012c)]. The corresponding results, concerning the functions (9.56), can be proved analogously using the specific asymptotic estimate (6.32). The Fatou type theorems for the series (9.50) and (9.51) are considered in [Paneva-Konovska (2014a, 2015a)], respectively. Their detailed proofs are set out in these publications, and follow the lines of those in Secs. 3.7 and 3.8 as well.

9.7. Overconvergence of Bessel Type Series

Let

$$\left\{\widetilde{J}_n^\mu(z)\right\}_{n=0}^\infty, \quad \left\{\widetilde{J}_{n-2\lambda,\lambda}^{\mu,q}(z)\right\}_{n=0}^\infty; \quad \mu > 0,$$

respectively

$$\left\{\widetilde{J}_{(\nu_{i_0}(n))}^{(m)}(z)\right\}_{n=0}^\infty, \quad \Re(\nu_i + 1) > 0 \quad (i = 1, 2, \ldots, m),$$

be the specified enumerable families of Bessel type functions, defined by (9.1) and (9.2) and the hyper-Bessel functions (9.48).

In this section, we consider the series in the above-discussed generalized Bessel functions of the kind (9.3), (9.4) and (9.51) in the complex plane. Namely, we consider the series:

$$\sum_{n=0}^\infty a_n \widetilde{J}_n^\mu(z), \quad \sum_{n=0}^\infty a_n \widetilde{J}_{n-2\lambda,\lambda}^{\mu,q}(z), \quad \sum_{n=0}^\infty a_n \widetilde{J}_{(\nu_{i_0}(n))}^{(m)}(z), \quad (9.61)$$

with complex coefficients, having Hadamard gaps, and study their overconvergence (in a sense of Remark 3.6). We only formulate the results, because their proofs follow those, referring to the Bessel series, given in Sec. 3.9.

Theorem 9.18 (about the overconvergence). *Let $\{a_n\}_{n=0}^\infty$ be a sequence of complex numbers satisfying the condition $\limsup_{n\to\infty} (|a_n|)^{1/n} = 1$, $\widetilde{f}(z)$ be the sum of any of the series (9.61) in the unit disk $D(0;1)$, $\widetilde{f}(z)$ have at least one regular point, belonging to the circle $C(0;1)$, and let $\widetilde{f}(z)$ possess Hadamard gaps. Then the corresponding series (9.61) is overconvergent.*

Theorem 9.19 (of Hadamard type about the gaps). *Let* $\{a_k\}_{k=0}^{\infty}$ *be a sequence of complex numbers satisfying the conditions as follows:* $\limsup_{n\to\infty} (|a_{k_n}|)^{1/k_n} = 1$ *for* $k_{n+1} \geq (1+\theta)k_n$ $(\theta > 0)$, $a_k = 0$ *for* $k_n < k < k_{n+1}$ *and* $\widetilde{f}(z)$ *be the sum of any of the series* (9.61) *in the unit disk* $D(0;1)$. *Then all the points of the unit circle* $C(0;1)$ *are singular for the function* $\widetilde{f}$, *i.e. the unit circle is a natural boundary of analyticity for the corresponding series* (9.61).

In conclusion, note that as particular cases one can obtain the corresponding results, referring to the Bessel series, which are given in Sec. 3.9, and they are also published in the paper [Paneva-Konovska (2015b)].

9.8. Overconvergence of Mittag-Leffler Type Series

Finally, let

$$\left\{\widetilde{E}_n(z)\right\}_{n=0}^{\infty}, \quad \left\{\widetilde{E}_{\alpha,n}^{\gamma}(z)\right\}_{n=0}^{\infty}; \quad \alpha, \gamma \in \mathbb{C}, \ \mathrm{Re}(\alpha) > 0,$$

respectively

$$\left\{\widetilde{E}_{(\alpha_i),(\mu_{i_0}(n))}^{(m)}(z)\right\}_{n=0}^{\infty}, \quad \mu_i \in \mathbb{C}, \ \alpha_i > 0 \ (i = 1, 2, \ldots, m),$$

be the families of Mittag-Leffler type functions, defined by (9.29) and multi-index Mittag-Leffler functions (9.47).

In this section, we consider the series in the above-discussed generalized Mittag-Leffler functions of the kind (9.30), (9.31) and (9.50) in the complex plane. Namely, we consider the series:

$$\sum_{n=0}^{\infty} a_n \widetilde{E}_n(z), \quad \sum_{n=0}^{\infty} a_n \widetilde{E}_{\alpha,n}^{\gamma}(z), \quad \sum_{n=0}^{\infty} a_n \widetilde{E}_{(\alpha_i),(\mu_{i_0}(n))}^{(m)}(z), \qquad (9.62)$$

with complex coefficients, having Hadamard gaps, and study their overconvergence (in a sense of Remark 3.6).

Again we only formulate the results, beginning with the overconvergence theorem. Since the proofs go analogously to those exposed in Sec. 3.9, they are omitted here.

Theorem 9.20 (about the overconvergence). *Let $\{a_n\}_{n=0}^{\infty}$ be a sequence of complex numbers satisfying the condition $\limsup_{n\to\infty} (|a_n|)^{1/n} = 1$, $\widetilde{f}(z)$ be the sum of any of the series (9.62) in the unit disk $D(0;1)$, $\widetilde{f}(z)$ have at least one regular point, belonging to the circle $C(0;1)$, and let $\widetilde{f}(z)$ possess Hadamard gaps. Then the corresponding series (9.62) is overconvergent.*

Theorem 9.21 (of Hadamard type about the gaps). *Let $\{a_k\}_{k=0}^{\infty}$ be a sequence of complex numbers satisfying the following conditions: $\limsup_{n\to\infty} (|a_{k_n}|)^{1/k_n} = 1$ for $k_{n+1} \geq (1+\theta)k_n$ $(\theta > 0)$, $a_k = 0$ for $k_n < k < k_{n+1}$ and $\widetilde{f}(z)$ be the sum of any of the series (9.62) in the unit disk $D(0;1)$. Then all the points of the unit circle $C(0;1)$ are singular for the function $\widetilde{f}$, i.e. the unit circle is a natural boundary of analyticity for the corresponding series (9.62).*

Before concluding this section, note that the results of Sec. 9.7, concerning the third series (9.61), could be obtained as corollaries of the corresponding results referring to the third series (9.62) as well.

9.9. Conclusions

In the preceding sections of this chapter and also in Chapter 3, we have considered the series in functions of different enumerable systems of the Bessel and Mittag-Leffler types, namely the systems (3.33), (9.1), (9.2), (9.29), (9.47) and (9.48). We have studied the convergence of such sorts of series in the complex plane, proving analogs to the classical results like Cauchy–Hadamard, Abel, Tauber, Littlewood and Fatou type theorems, as well as overconvergence theorems.

For convenience, let us briefly denote by

$$\widetilde{J} := \{\widetilde{j}_n\}_{n\in\mathbb{N}_0}$$

any of the families listed above. In the process of studying the series

$$\sum_{n=0}^{\infty} a_n \widetilde{j}_n(z), \quad a_n \in \mathbb{C}, \ z \in \mathbb{C}, \tag{9.63}$$

we also consider along with it the power series $\sum_{n=0}^{\infty} a_n z^n$ with the same coefficients a_n and $z \in \mathbb{C}$.

We emphasize that the results obtained for the series (9.63) are quite analogous to the ones for the classical power series.

As has been well seen, these series have the same radius of convergence R, and they are both absolutely and uniformly convergent on each closed disk $|z| \leq r$ $(r < R)$. Moreover, if each of them converges at the point z_0 of the boundary of $D(0; R)$, then the theorems of Abel type hold for both series in the same angular region. Further, if $R = 1$ and $\{a_n\}_{n=0}^{\infty}$ is a sequence of complex numbers satisfying the condition $\lim_{n \to \infty} a_n = 0$, and all the points (including the ends) of the arc γ of the unit circle $C(0; 1)$ are regular to the sums of both considered series, then the series (9.63) and $\sum_{n=0}^{\infty} a_n z^n$ converge even uniformly, on the arc γ. Moreover, if additionally the sums $\widetilde{f}$ of both series have Hadamard gaps, as well as at least one regular point belonging to the circle $C(0; 1)$, then they are overconvergent. However, if also $\limsup_{n \to \infty} (|a_{k_n}|)^{1/k_n} = 1$ for $k_{n+1} \geq (1 + \theta)k_n$ $(\theta > 0)$ and $a_k = 0$ for $k_n < k < k_{n+1}$, then all the points of the unit circle $C(0; 1)$ are singular for the function $\widetilde{f}$, i.e. the unit circle is a natural boundary of analyticity for the corresponding series.

For further considerations, by letting $z_0 \in C(0; R)$ and $\widetilde{j}_n(z_0) \neq 0$, we normalize in some sense the system (9.63) and denote

$$(J; z_0) := \{j_n\}_{n \in \mathbb{N}_0}, \quad j_n(z) = j_n^*(z; z_0) = \frac{\widetilde{j}_n(z)}{\widetilde{j}_n(z_0)}. \tag{9.64}$$

It is evident that all the functions j_n in formula (9.64) are entire functions satisfying the condition $j_n(z_0) = 1$, i.e. the corresponding family is of the kind (3.22). In order to provide the further results, we come back to the Definition 3.6 of (J, z_0)-summation of a given numerical series. As a matter of fact, if the numerical series (3.7) is (J, z_0)-summable, by means of the family (9.64), and additionally $\lim_{n \to \infty} n a_n = 0$ (or even $a_n = O(1/n)$), then it is convergent. This means that Tauber and Littlewood type theorems hold, analogously to the power series theory. Surely, this summation is regular and its regularity follows from the existing Abel type theorems.

Finally, the results obtained in this chapter, and also in Chapter 3 (as particular cases), can be briefly summarized in the following way. The basic properties of the series, objects of this survey, are sufficiently 'close' to those of the power series, i.e. their behaviors are quite similar to the one of the widely used power series.

Bibliography

Agarwal, R. P. (1953). A propos d'une note de M. Pierre Humbert. *C.R. Seances Acad. Sci.*, **236**(21), 2031–2032.

Askey, R. (1975). *Orthogonal Polynomials and Special Functions*. Philadelphia, PA: SIAM.

Baleanu, D., Diethelm, K., Scalas, E. and Trujillo, J. J. (2012). *Fractional Calculus: Models and Numerical Methods*, Series on Complexity, Nonlinearity and Chaos. Singapore: World Scientific.

Bazhlekova, E. and Dimovski, I. (2013). Exact solution for the fractional cable equation with nonlocal boundary conditions. *Cent. Eur. J. Phys.*, **11**(10), 1304–1313, doi:10.2478/s11534-013-0213-5.

Bazhlekova, E. and Dimovski, I. (2014). Exact solution of two-term time-fractional Thornleys problem by operational method. *Integr. Transf. Spec. Funct.*, **25**(1), 61–74, doi:10.1080/10652469.2013.815184.

Bernstein, Vl. (1933). *Leçons sur les Progres Recents de la Theorie des Series de Dirichlet*. Paris: Gauthier-Villars.

Bieberbach, L. (1955). *Analytische Fortsetzung*. Berlin: Springer-Verlag.

Boyadjiev, L. I. (1986). Abel's Theorems for Laguerre and Hermite Series, *C.R. Acad. Bulg. Sci.*, **39**(4), 13–15.

Boyadjiev, L. (1990). Singular points and series in Laguerre polynomials (in Bulgarian). In *Proc. Jubillee Session Devoted to 100th Birthday of Acad. Chakalov*, Samokov, Bulgaria (1986), pp. 40–45.

Cowling, V. F. (1958). Series of Legendre and Laguerre Polynomials. *Duke Math. J.* **25**(1), 171–176.

Delerue, P. (1953). Sur le calcul symbolique à n variables et fonctions hyper-besséliennes (II). *Ann. Soc. Sci. Brux. I*, **67**(3), 229–274.

Dimovski, I. (1966). Operational calculus for a class of differential operators. *C.R. Acad. Bulg. Sci.*, **19**(12), 1111–1114.

Dimovski, I. and Kiryakova, V. (1986). Generalized Poisson transmutations and corresponding representations of hyper-Bessel functions. *C.R. Acad. Bulg. Sci.*, **39**(10), 29–32.

Dimovski, I. and Kiryakova, V. (1987). Generalized Poisson representations of hypergeometric functions $_pF_q$, $p < q$, using fractional integrals. In *Proc. 16th Spring Conf. Union of Bulgarian Mathematicians*, Sofia, Bulgaria (1987), pp. 205–212.

Dzrbashjan, M. M. (1960). On the integral transformation generated by the generalized Mittag-Leffler function (in Russian). *Izv. Akad. Nauk Arm. SSR*, **13**(3), 21–63.

Dzrbashjan, M. M. (1966). *Integral Transforms and Representations in the Complex Domain* (in Russian). Moscow: Nauka.

Erdélyi, A. *et al.* (eds.) (1953). *Higher Transcendental Functions*, Vols. 1 and 2. New York: McGraw-Hill.

Erdélyi, A. *et al.* (eds.) (1955). *Higher Transcendental Functions*, Vol. 3. New York: McGraw-Hill.

Euler, L. (1787). *Institutiones Calculi Differentialis*, Vol. 1. Ticini: P. Galeatus.

Gajić, L. and Stanković, B. (1976). Some properties of Wright's function. *Publ. l'Inst. Math. Beograd, Nouv. Ser.*, **20**(34), 91–98.

Gorenflo, R., Kilbas, A. A., Mainardi, F. and Rogosin, S. V. (2014). *Mittag-Leffler Functions: Related Topics and Applications*. Berlin: Springer-Verlag.

Gorenflo, R. and Mainardi, F. (2000). On Mittag-Leffler-type functions in functional evolution processes. *J. Comput. Appl. Math.*, **118**(1–2), 283–299.

Hadamard, J. (1892). Essai sur l'etude des fonctions données par leur développement de Taylor. *J. Math. Pures Appl. Ser. IV*, **8**, 101–186.

Hadamard, J. (1893). Etude sur les propriétés des fonctions entières et en particulier d'une fonction considérée par Riemann. *J. Math. Pures Appl. Ser. IV*, **9**, 171–215.

Hardy, G. (1949). *Divergent Series*. Oxford: Oxford University Press.

Herzallah, M. A. E. and Baleanu, D. (2012). Existence of a periodic mild solution for a nonlinear fractional differential equation. *Comput. Math. Appl.*, **64**(10), 3059–3064.

Hilfer, R. (ed.) (2000). *Applications of Fractional Calculus in Physics*. Singapore: World Scientific.

Hille, E. (1939). Contribution to the theory of Hermitian series (I). *Duke Math. J.* **5**, 875–936.

Hille, E. (1940). Contributions to the theory of Hermitian series II: The representation problem. *Trans. Am. Math. Soc.*, **47**(1), 80–94.

Hille, E. and Tamarkin, J. D. (1930). On the theory of linear integral equations. *Ann. Math.*, **31**, 479–528.

Humbert, P. and Agarwal, R. P. (1953). Sur la fonction de Mittag-Leffler et quelquesunes de ses généralisations. *Bull. Sci. Math.*, **77**(10), 180–185.

Kazmin, V. (1960). On the subsequences of Hermite and Laguerre polynomials (in Russian). *Vestn. Mosk. Univ. 1, Mat. Mekh.*, **2**, 6–9.

Kilbas, A. A. and Koroleva, A. A. (2005). Generalized Mittag-Leffler function and its extension (in Russian). *Tr. Inst. Mat. Minsk*, **13**(1), 43–52.

Kilbas, A. A. and Koroleva, A. A. (2006). Integral transform with the extended generalized Mittag-Leffler function. *Math. Model. Anal.*, **11**(2), 173–186.

Kilbas, A. A., Koroleva, A. A. and Rogosin, S. V. (2013). Multi-parametric Mittag-Leffler functions and their extension. *Fract. Calc. Appl. Anal.*, **16**(2), 378–404, doi:10.2478/s13540-013-0024-9.

Kilbas, A. A. and Saigo, M. (2004). *H-Transform: Theory and Applications*. Boca Raton: Chapman and Hall/CRC.

Kilbas, A. A., Saigo, M. and Saxena, R. K. (2004). Generalized Mittag-Leffler function and generalized fractional calculus operators. *Integr. Transf. Spec. Funct.*, **15**(1), 31–49.

Kilbas, A. A., Srivastava, H. M. and Trujillo, J. J. (2006). *Theory and Applications of Fractional Differential Equations*. North-Holland Mathematical Studies. New York, NY: Elsevier Science.

Kiryakova, V. (1994). *Generalized Fractional Calculus and Applications*. New York: Longman Scientific and Technical/John Wiley and Sons.

Kiryakova, V. (1997). All the special functions are fractional differintegrals of elementary functions. *J. Phys. A, Math. Gen.*, **30**(14), 5085–5103, doi:10.1088/0305-4470/30/14/019.

Kiryakova, V. (1999). Multiindex Mittag-Leffler functions, related Gelfond-Leontiev operators and Laplace type integral transforms. *Fract. Calc. Appl. Anal.*, **2**(4), 445–462.

Kiryakova, V. (2000). Multiple (multiindex) Mittag-Leffler functions and relations to generalized fractional calculus. *J. Comput. Appl. Math.*, **118**, 241–259, doi:10.1016/S0377-0427(00)00292-2.

Kiryakova, V. S. (2008). Some special functions related to fractional calculus and fractional (non-integer) order control systems and equations. *Facta Univ., Ser.: Autom. Control Robot.*, **7**(1), 79–98.

Kiryakova, V. (2010a). The multi-index Mittag-Leffler functions as important class of special functions of fractional calculus. *Comput. Math. Appl.*, **59**(5), 1885–1895, doi:10.1016/j.camwa.2009.08.025.

Kiryakova, V. (2010b). The special functions of fractional calculus as generalized fractional calculus operators of some basic functions. *Comput. Math. Appl.*, **59**(3), 1128–1141, doi:10.1016/j.camwa.2009.05.014.

Kiryakova, V. (2011). Fractional order differential and integral equations with Erdélyi–Kober operators: Explicit solutions by means of the transmutation method. *AIP Conf. Proc.*, **1410**, 247–258, doi:10.1063/ 1.3664376.

Kiryakova, V. (2014). From the hyper-Bessel operators of Dimovski to the generalized fractional calculus. *Fract. Calc. Appl. Anal.*, **17**(4), 977–1000, doi:10.2478/s13540-014-0210-4.

Kiryakova, V. and Hernandez-Suarez, V. (1995). Bessel–Clifford third order differential operator and corresponding Laplace type integral transform. *Diss. Math.*, **340**, 143–161.

Kiryakova, V. and Luchko, Yu. (2010). The multi-index Mittag-Leffler functions and their applications for solving fractional order problems in applied analysis. *AIP Conf. Proc.*, **1301**, 597–613, Sozopol, Bulgaria, Vol. 1301 (2010), doi:10.1063/1.3526661.

Kljuchantzev, M. (1975). On the construction of r-even solutions of singular differential equations (in Russian). *Dokl. Akad. Nauk. SSSR*, **224**(5), 1000–1008.

Kljuchantzev, M. (1987). An introduction to the theory of $(\nu_1, \ldots, \nu_{r-1})$-transforms (in Russian). *Mat. Sb.*, **132**(2), 167–181.

Korevaar, J. (2002). A century of complex Tauberian theory. *Bull. (New Ser.) Am. Math. Soc.*, **39**(4), 475–531, doi:10.1090/S0273-0979(02)00951-5.

Kovacheva, R. (2008). Zeros of sequences of partial sums and overconvergence. *Serdica Math. J.*, **34**, 467–482.

Lehua, M. (1996). On Neumann–Bessel series, *Approx. Theor. Appl.*, **12**(1), 68–77.

Leont'ev, A. (1980). *Sequences of Exponential Polynomials* (in Russian). Moscow: Nauka.

Levin, B. (1964). *Distribution of Zeros of Entire Functions*. Providence: American Mathematical Society.

Luchko, Y. (1999). Operational method in fractional calculus. *Fract. Calc. Appl. Anal.*, **2**(4), 463–488.

Luchko, Y. and Gorenflo, R. (1998). Scale invariant solutions of a PDE of fractional order. *Fract. Calc. Appl. Anal.*, **1**(1), 63–78.

Luchko, Y. and Gorenflo, R. (1999). An operational method for solving fractional differential equations with the Caputo derivatives. *Acta Math. Vietnam.*, **24**(2), 207–233.

Mainardi, F. (2010). *Fractional Calculus and Waves in Linear Viscoelasticity: An Introduction to Mathematical Models*. London: Imperial College Press World Scientific.

Mandelbrojt, S. (1927). *Modern Researches on the Singularities of Functions Defined by Taylor's Series*. The Rice Institute Pamphlet 4, Vol. 14. Houston, TX: Rice University.

Marichev, O. I. (1978). *A Method of Calculating Integrals of Special Functions (Theory and Tables of Formulas)* [In Russian: *Metod vychisleniya integralov ot spetsial'nykh funktsij (Teoriya i tablitsy formul)*]. Minsk: Nauka i Technika.

Markushevich, A. (1967). *A Theory of Analytic Functions*, Vols. 1, and 2 (in Russian). Moscow: Nauka.

Mastroianni, G. and Milovanovic, G. V. (2008). *Interpolation Processes — Basic Theory and Applications*. Berlin: Springer-Verlag.

Mathai, A. M. and Haubold, H. J. (2008). *Special Functions for Applied Scientists*. New York: Springer-Verlag.

Mathai, A. M. and Haubold, H. J. (2011). Matrix-variate statistical distributions and fractional calculus. *Fract. Calc. Appl. Anal.*, **14**, 138–157, doi:10.2478/s13540-011-0010-z.

Mathai, A. M. and Saxena, R. K. (1978). *The H-function with Applications in Statistics and Other Disciplines*. New Delhi, Wiley Eastern.

Mathai, A. M., Saxena, R. K. and Haubold, H. J. (2010). *The H-function: Theory and Applications*. New York: Springer-Verlag.

Milovanovic, G. V., Mitrinovic, D. S. and Rassias, Th. M. (1994). *Topics in Polynomials: Extremal Problems, Inequalities, Zeros*. Singapore: World Scientific.

Mittag-Leffler, M. G. (1899). Sur la représentation analytique d'une branche uniforme d'une fonction monogène (première note). *Acta Math.*, **23**, 43–62.

Mittag-Leffler, M. G. (1900a). Sur la représentation analytique d'une branche uniforme d'une fonction monogéne (seconde note). *Acta Math.*, **24**, 183–204.

Mittag-Leffler, M. G. (1900b). Sur la représentation analytique d'une branche uniforme d'une fonction monogéne (troiséme note). *Acta Math.*, **24**, 205–244.

Mittag-Leffler, M. G. (1902). Sur la représentation analytique d'une branche uniforme d'une fonction monogéne (quatriéme note). *Acta Math.*, **26**, 353–392.

Mittag-Leffler, M. G. (1905). Sur la représentation analytique d'une branche uniforme d'une fonction monogéne (cinquiéme note). *Acta Math.*, **29**, 101–181.

Mittag-Leffler, M. G. (1920). Sur la représentation analytique d'une branche uniforme d'une fonction monogéne (sixiéme note). *Acta Math.*, **42**, 285–308.

Nehari, Z. (1956). On the singularities of Legendre expansions. *Arch. Ration. Mech. Anal.*, **5**(6), 987–992.

Nigmatullin, R. R. and Baleanu, D. (2012). The derivation of the generalized functional equations describing self-similar processes. *Fract. Calc. Appl. Anal.*, **15**(4), 718–740, doi:10.2478/s13540-012-0049-5.

Ostrowski, A. (1921). Über eine Eigenschaft gewisser Potenzreihen mit unendlich vielen verschwindenden Koefizienten. *Sitz.ber., Preuss. Akad. Wiss.*, **34**, 557–565.

Ostrowski, A. (1923). Über allgemeine Konvergenzsätze der komplexen Funktionentheorie. Jahresber. *Dtsch. Math.-Ver.*, **32**, 185–194.

Pagnini, G. (2012). Erdélyi–Kober fractional diffusion. *Fract. Calc. Appl. Anal.*, **15**(1), 117–127, doi:10.2478/s13540-012-0008-1.

Paneva-Konovska, J. (1999). A Tauber type theorem for series in Bessel functions. *Fract. Calc. Appl. Anal.*, **2**(5), 683–688.

Paneva-Konovska, J. (2003). Complete systems of Bessel functions of first kind in spaces of holomorphic functions. *Fract. Calc. Appl. Anal.* **6**(3), 299–309.

Paneva-Konovska, J. (2007). Index-asymptotic formulae for Wright's generalized Bessel functions. *Math. Sci. Res. J.* **11**(5), 424–431.

Paneva-Konovska, J. (2008). Tauberian theorem of Littlewood type for series in Bessel–Maitland functions. *Math. Maced.*, **6**, 55–60.

Paneva-Konovska, J. (2009). Tauberian theorem of Littlewood type for series in Bessel functions of first kind. *C.R. Acad. Bulg. Sci.*, **62**(2), 161–166.

Paneva-Konovska, J. (2010a). Convergence of series in Mittag-Leffler functions. *C.R. Acad. Bulg. Sci.*, **63**(6), 815–822.

Paneva-Konovska, J. (2010b). Inequalities and asymptotic formulae related to generalizations of the Bessel functions. *AIP Conf. Proc.*, **1293**, 157–164, doi:10.1063/1.3515580.

Paneva-Konovska, J. (2010c). Series in Mittag-Leffler functions: Inequalities and convergent theorems. *Fract. Calc. Appl. Anal.*, **13**(4), 403–414.

Paneva-Konovska, J. (2011). Multi-index ($3m$-parametric) Mittag-Leffler functions and fractional calculus. *C.R. Acad. Bulg. Sci.*, **64**(8), 1089–1098.

Paneva-Konovska, J. (2012a). Inequalities and asymptotic formulae for the three-parametric Mittag-Leffler functions. *Math. Balkan., New Ser.*, **26**(1–2), 203–210.

Paneva-Konovska, J. (2012b). Series in Mittag-Leffler functions: Geometry of convergence. *Adv. Math. Sci. J.*, **1**(2), 73–79.

Paneva-Konovska, J. (2012c). The convergence of series in multi-index Mittag-Leffler functions. *Integr. Transf. Spec. Funct.*, **23**(3), 207–221, doi:10.1080/10652469.2011.575567.

Paneva-Konovska, J. (2013a). Comparison between the convergence of power and Bessel series. *AIP Conf. Proc.*, **1570**, 383–392, doi:10.1063/1.4854780.

Paneva-Konovska, J. (2013b). On the multi-index (3m-parametric) Mittag-Leffler functions, fractional calculus relations and series convergence. *Cent. Eur. J. Phys.*, **11**(10), 1164–1177, doi:10.2478/s11534-013-0263-8.

Paneva-Konovska, J. (2014a). A family of hyper-Bessel functions and convergent series in them. *Fract. Calc. Appl. Anal.*, **17**(4), 1001–1015, doi:10.2478/s13540-014-0211-3.

Paneva-Konovska, J. (2014b). Convergence of series in three-parametric Mittag-Leffler functions. *Math. Slovaca*, **64**(1), 73–84, doi:10.2478/s12175-013-0188-0.

Paneva-Konovska, J. (2014c). Fatou theorems for multi-index Bessel series. *AIP Conf. Proc.*, **1631**, 303–312, doi:10.1063/1.4902491.

Paneva-Konovska, J. (2014d). Series in Prabhakar functions and the geometry of their convergence. In *Mathematics in Industry*, A. Slavova (ed.), Chap. 5, pp. 198–214. Cambridge, UK: Cambridge Scholars Publishing.

Paneva-Konovska, J. (2015a). On the multi-index Mittag-Leffler series and their boundary behaviour. *Integr. Transf. Spec. Funct.*, **26**(3), 152–164, doi:10.1080/10652469.2014.975129.

Paneva-Konovska, J. (2015b). Overconvergence of Bessel series. *C.R. Acad. Bulg. Sci.*, **68**(9), 1091–1098.

Pathak, R. S. (1966). Certain convergence theorems and asymptotic properties of a generalization of Lommel and Maitland transformations. *Proc. Natl. Acad. Sci. India A*, **36**(1), 81–86.

Podlubny, I. (1999). *Fractional Differential Equations*. San Diego, CA: Academic Press.

Pólya, G. (1927). Eine, Verallgemeinerung des Fabryschen. *Nachr. Ges. Wiss. Gottingen. Math.-naturwiss. Kl.*, **1927**, 187–195.

Pólya, G. (1929). Untersuchungen über Lücken und Singularitäten von Potenzreihen. *Math. Z.*, **29**(1), 549–640.

Pólya, G. and Szego, G. (1925). *Aufgaben und Lehrsätze aus der Analysis*, Vols. I, and II. Berlin: Springer-Verlag.

Prabhakar, T. R. (1971). A singular integral equation with a generalized Mittag-Leffler function in the kernel. *Yokohama Math. J.*, **19**, 7–15.

Prieto, A. I., de Romero, S. S. and Srivastava, H. M. (2007). Some fractional-calculus results involving the generalized Lommel–Wright and related functions, *Appl. Math. Lett.*, **20**, 17–22.

Prudnikov, A. A., Brychkov, Yu. A. and Marichev, O. I. (1990). *Integrals and Series. More Special Functions*, Vol. 3. New York, NY: Gordon and Breach Science Publishers.

Rajkovic, P., Marinkovic, S. and Stankovic, M. (2008). *Diferencijalno-Integralni Racun Baicnih Hipergeometrijskih Funkcija* (in Serbian). Serbia: Mašinski Fakultet, Univerzitet u Nišu.

Rusev, P. (1977a). A theorem of a Tauber type for summation by means Laguerre polynomials (in Russian). *C.R. Acad. Bulg. Sci.*, **30**(3), 331–334.

Rusev, P. (1977b). About the completeness of the system of Laguerre functions of second kind (in Russian). *C.R. Acad. Bulg. Sci.*, **33**(1), 9–11.

Rusev, P. (1980). Completeness of Laguerre and Hermite functions of second kind. In *Proc. Constructive Function Theory '77*, Sofia, Bulgaria (1980), pp. 469–473.

Rusev, P. (1984). *Analytic Functions and Classical Orthogonal Polynomials.* Sofia: Publishing House of the Bulgarian Academy of Sciences.

Rusev, P. (1994a). Complete systems of Jacobi associated functions in spaces of holomorphic functions. *Analysis*, **14**, 249–255.

Rusev, P. (1994b). Complete systems of Tricomi functions in spaces of holomorphic functions. *Annu. Univ. Sofia Fac. Math. Inform.*, **88**, 1–6.

Rusev, P. (1995). Complete systems of Kummer and Weber–Hermite functions in spaces of holomorphic functions. *Symposia Gaussiana, Conference A: Mathematics and Theoretical Physics*, Part I, 723–731. Berlin: De Gruyter.

Rusev, P. (1997). Completeness of systems related to Laguerre and Hermite polynomials and associated functions. *Analysis*, **17**, 323–334.

Rusev, P. (2005). *Classical Orthogonal Polynomials and Their Associated Functions in Complex Domain*, Bulgarian Acadamic Monograph 10. Sofia: Marin Drinov Academic Publishing House.

Samko, S., Kilbas, A. and Marichev, O. (1993). *Fractional Integrals and Derivatives: Theory and Applications.* Yverdon, Switzerland: Gordon and Breach Science Publishers.

Sandev, T., Chechkin, A., Kantz, H. and Metzler, R. (2015). Diffusion and Fokker–Planck–Smoluchowski equations with generalized memory kernel. *Fract. Calc. Appl. Anal.*, **18**(4), 1006–1038, doi:10.1515/fca-2015-0059.

Sandev, T., Metzler, R. and Tomovski, Ž. (2012). Velocity and displacement correlation functions for fractional generalized Langevin equations.

Fract. Calc. Appl. Anal., **15**(3), 426–450, doi:10.2478/s13540-012-0031-2.

Sandev, T., Metzler, R. and Tomovski, Ž. (2014). Correlation functions for the fractional generalized Langevin equation in the presence of internal and external noise. *J. Math. Phys.*, **55**, 023301, doi:10.1063/1.4863478.

Sandev, T., Tomovski, Ž. and Dubbeldam, J. (2011). Generalized Langevin equation with a three parameter Mittag-Leffler noise. *Physica A*, **390**, 3627–3636, doi:10.1016/j.physa.2011.05.039.

Schäfke, F. W. (1960). Reihenentwicklungen analytischer Funktionen nach Biorthogonalsystemen spezieller Funktionen. I. *Math. Z.*, **74**, 436–470.

Schäfke, F. W. (1961). Reihenentwicklungen analytischer Funktionen nach Biorthogonalsystemen spezieller Funktionen. II. *Math. Z.*, **75**, 154–191.

Schäfke, F. W. (1963). Reihenentwicklungen analytischer Funktionen nach Biorthogonalsystemen spezieller Funktionen. III. *Math. Z.*, **80**, 400–442.

Tchakalov, L. (1972). *Introduction in the Theory of Analytic Functions* (in Bulgarian) Sofia: Nauka i Izkustvo.

Titchmarsh, E. C. (1986). *Introduction to the Theory of Fourier Integrals*, 3rd edn. New York: Chelsea Publishing [First edn. published in 1937 by Oxford University Press].

Watson, G. N. (1949). *Theory of Bessel Functions*, Vol. 1 (in Russian) Moscow: Inostrannaya Literatura.

Whittaker, E. T. and Watson, G. N. (1963). *A Course of Modern Analysis*, Vols. 1 and 2 (in Russian). Moscow: State Publishing House for Physics-Mathematics Literature.

Wiman, A. (1905). Über den Fundamentalsatz in der Theorie der Funktionen $E_\alpha(x)$. *Acta Math.*, **29**, 191–201.

Wright, E. M. (1933). On the coefficients of power series having exponential singularities. *J. London Math. Soc.*, **8**, 71–79.

Yakubovich, S. and Luchko, Yu. (1994). *The Hypergeometric Approach to Integral Transforms and Convolutions*, 1st edn. Dordrecht: Kluwer Academic.

Zygmund, A. and Saks, S. (1952). *Analytic Functions* Warszawa, Poland: Instytut Matematyczny Polskiej Akademi Nauk.

Index

Printed in the United States
By Bookmasters